The United Kingdom's Naval Shipbuilding Industrial Base

The Next Fifteen Years

Mark V. Arena

Hans Pung

Cynthia R. Cook

Jefferson P. Marquis

Jessie Riposo

Gordon T. Lee

Prepared for the United Kingdom's Ministry of Defence

Approved for public release; distribution unlimited

The research described in this report was sponsored by the United Kingdom's Ministry of Defence. The research was conducted jointly in RAND Europe and the RAND National Security Research Division.

Library of Congress Cataloging-in-Publication Data is available for this publicaton.

ISBN 0-8330-3706-4

The RAND Corporation is a nonprofit research organization providing objective analysis and effective solutions that address the challenges facing the public and private sectors around the world. RAND's publications do not necessarily reflect the opinions of its research clients and sponsors.

RAND® is a registered trademark.

Cover design by Barbara Angell Caslon

Cover photo from Reuters, via Landov LLC

© Copyright 2005 RAND Corporation

All rights reserved. No part of this book may be reproduced in any form by any electronic or mechanical means (including photocopying, recording, or information storage and retrieval) without permission in writing from RAND.

Published 2005 by the RAND Corporation
1776 Main Street, P.O. Box 2138, Santa Monica, CA 90407-2138
1200 South Hayes Street, Arlington, VA 22202-5050
201 North Craig Street, Suite 202, Pittsburgh, PA 15213-1516
RAND URL: http://www.rand.org/
To order RAND documents or to obtain additional information, contact
Distribution Services: Telephone: (310) 451-7002;
Fax: (310) 451-6915; Email: order@rand.org

Preface

In the autumn of 2003, the United Kingdom's Ministry of Defence (MOD) engaged the RAND Corporation to study the domestic capacity for naval ship construction. The impetus for the study was a concern on the MOD's part that the confluence of several shipbuilding programmes (i.e., Astute, MARS, CVF, FSC, JCTS, and Type 45[1]) could potentially overburden the industry. The objective of the study was to take a strategic look at the shipbuilding industry over the next 15 years to determine where there might be capacity limitations and to offer recommendations as to how any identified limitations might be addressed. For example, are there production skills or trades that will be in short supply? If so, what policy options are open to the government to remedy such a shortfall (training incentives, shifting of work, etc.)? The scope of the study was limited to the UK industry, in line with current defence procurement policy. This report is the final product of that study and summarises the analysis.

We organised our analysis by decomposing capacity into three major elements: labour, facilities, and suppliers. Labour encompassed all aspects of ship production (manufacture, design, engineering, management, outfitting, and support). The facilities analysis addressed the throughput limitations of the major shipyard assets, such as piers, docks, and slipways. For simplicity, we limited this examination to facilities involved in final assembly and afloat outfitting. The

---

[1] For a full listing and description of these ships, see Table S.1 in the Summary.

suppliers make up a major portion of the shipbuilding value chain and provide a wide range of different products and services—from painting services to complex weapon systems. The suppliers' ability to meet any peak in demand will affect the ability of the MOD to procure ships within the desired time frame and budget.

For our capacity evaluations, we relied on data surveys and interviews with many firms and organisations associated with shipbuilding in the United Kingdom, including shipbuilders, ship repairers, suppliers, industry associations, and government organisations. This interaction took the better part of five months.

This report should be of special interest not only to the MOD's Defence Procurement Agency (DPA) but also to service and defence agency managers and policymakers involved in weapon system acquisitions on both sides of the Atlantic. It should also be of interest to shipbuilding industrial executives in the United Kingdom.

This research was sponsored by the MOD and conducted within RAND Europe and the International Security and Defense Policy Center of the RAND National Security Research Division, which conducts research for the US Department of Defense, allied foreign governments, the intelligence community, and foundations.

For more information on RAND Europe, contact the president, Martin van der Mandele. He can be reached by email at mandele@rand.org; by phone at +31 71 524 5151; or by mail at RAND Europe, Netonweg 1, 2333 CP Leiden, The Netherlands. For more information on the International Security and Defense Policy Center, contact the director, Jim Dobbins. He can be reached by email at James_Dobbins@rand.org; by phone at (310) 393-0411, extension 5134; or by mail at RAND Corporation, 1200 South Hayes Street, Arlington, VA 22202-5050 USA. More information about RAND is available at www.rand.org.

Contents

Preface iii
Figures xi
Tables xv
Summary xvii
Acknowledgements xxxv
Abbreviations xxxvii

CHAPTER ONE
Introduction 1
Warship Production Is a Unique Industry 2
MOD Ship Programmes 5
    Programmes Past Main Gate 6
    Projects Pre–Main Gate 7
    Other Speculative Programmes 9
    Notional Programme Timelines 10
Business Environment of UK Naval Shipbuilding Industrial Base
    Since 1985 12
Issues for Policymakers 16
Study Structure 17
    Plan 18
    Labour 19
    Facilities 19
    Suppliers 19
Survey of the UK Shipbuilding Industry 20
Study Outline 21

CHAPTER TWO
Labour Demand 23
Methodology 24
Basic Assumptions 27
  Additional Assumptions 31
Current MOD Plan: Overall Labour Demand 33
Current MOD Plan: Demand for Specific Labour Skills 35
  Management Labour Skills 35
  Technical Labour Skills 36
  Structural Labour Skills 37
  Outfitting Labour Skills 38
  Support Labour Skills 39
Macro Versus Micro View of Demand 39
Alternate Future Scenarios 41
  Scenario 1: Decreased MOD Requirements or Budgets 42
  Scenario 2: Addition of Future Submarine to the MOD's
    Requirements 44
  Scenario 3: Increased MOD Future Requirements 47
  Future MOD Programme Challenges 50
Options for Managing Increased MOD Demand 51
  Illustrative Results of Level-Loading Future MOD
    Labour Demand 55
  Other Build Strategies 58
Summary 59

CHAPTER THREE
The Supply of Naval Shipyard Labour in the United Kingdom 61
Employment Status of the UK Shipbuilding and Repair
  Industrial Base 62
  Regional Differences in UK Shipyard-Related Employment 62
  Sector Employment in the UK Shipbuilding and Repair Industry 63
  UK Shipbuilding and Repair Industry Workers Are Ageing 66
  Small Reliance on Temporary Workers 67
Ability of the Naval Shipyards to Expand Their Workforces 68
  Concerns About Labour Shortages 68

Recruitment in the Shipbuilding and Repair Industry Faces Significant
    Obstacles 69
Shipyard Training Initiatives 70
Consequences of Unemployment, Demographic Changes, and Shipyard
    Redundancies 71
Recent Shipyard Recruiting Efforts 73
Pools of Labour That Could Be Tapped 75
Shipyards Rely on Outsourcing to Varying Degrees 76
A Comparison of the Supply of Naval Workers with the Demand Under
    Different Future Scenarios 77
    Three Supply Cases 78
    Shipyard Labour Supply Model 80
    Results of the Shipyard Supply Analysis 81
Concluding Observations 86

CHAPTER FOUR
Facilities Utilisation at the UK Shipyards 89
Ship Production Facilities and Phases 89
How We Studied Facilities and Phases 91
    Identifying Demand and Assigning Facilities to Phases 92
Final Assembly Facilities' Capacity and Considerations 93
Afloat Outfitting Facilities Capacity Considerations 96
Capacity Implications for Future Programmes 100
    Type 45 101
    CVF 106
    MARS 107
    Astute 108
    LSD(A) 109
    Future Surface Combatant 109
    Joint Casualty Treatment Ship 110
Summary 110

CHAPTER FIVE
The UK Shipbuilding Supplier Industrial Base 113
Research Approach 113
Characterising the Supplier Base—The Shipyard Perspective 115

What They Supply 115
Where They Are 117
Three Measures of Supplier Strength 118
Summary 122
Supplier Survey Results 122
Demographic Information on Sample Suppliers 123
Suppliers' Business Base 124
Number of Customers 127
Number of Competitors 128
Recruiting Challenges 129
Engineers Presented the Most Challenges for Recruiting 129
Challenges Working for the MOD 130
Summary 132
Results from Linking Shipyard and Supplier Surveys 132
Developing an Effective Supplier Strategy 133
Conclusion 134

CHAPTER SIX
Nontraditional Sources for Naval Shipbuilding:
  Commercial Shipbuilding and Offshore Industries 137
Declining Markets for Offshore and Commercial Work 138
Potential Resources Available 140
    Labour 141
    Facilities 143
Strengths and Weaknesses of Using Offshore Firms in Naval
    Production 143
Summary 146

CHAPTER SEVEN
Issues for the Ministry of Defence to Consider 147
Summary 147
    Labour Demand 147
    Labour Supply 149
    Facilities 150
    Suppliers 151
Potential Remedial Actions That MOD Can Take 151

    The MOD Needs to Make Long-Term Industrial Planning Part of
        the Acquisition Process 154
    Define the Appropriate Role of the Offshore Industry 156
    Carefully Consider the Implications of Foreign Procurement of
        Complete Ships 157
    Labour Wage Pressures During Peak Demand 158
    Encourage Long-Term Investment Through Multi-Ship Contracts .. 159
    Consider the Feasibility of Competition in Light of the Industrial Base
        Constraints 159
    Explore the Advantages of Common Design Tools 160
Conclusions 160

APPENDIX
A. Effects of Schedule Slippage on MOD Labour Demands 163
B. Ship Dimensions 171
C. Skill Breakout, by Management/Technical and Manufacturing
    Categories 173

References 175

Figures

S.1. Schedule of MOD Naval Programmes, 2005–2020 xx
S.2. Future MOD Labour Demands, by Programme xxiv
S.3. Direct Labour Demands for the Current Plan and a Level-Loaded Example xxv
S.4. Projected Shipyard Labour Supply Without Additional Recruitment, 2004–2020 xxvi
S.5. Shipyard Employment Projections Versus Demand xxviii
S.6. Total UK Final Assembly and Afloat Outfitting Facility Requirements, 2004–2020 xxix
1.1. Gantt Chart of MOD Naval Programmes over the Next 15 Years 10
1.2. Royal Navy Fleet Size over the Past Four Decades 12
1.3. Number of Combatants Delivered Each Year 13
1.4. History of Naval Shipbuilders Post-Privatisation 15
1.5. Study Hierarchy 18
2.1. RAND's Basic Labour Forecasting Model 24
2.2. Example of Direct Labour Distribution Curves for an Individual Ship Class 27
2.3. Individual Ship Aggregation to Represent Entire Shipbuilding Programme 28
2.4. Future MOD Labour Demand, by Programme 33
2.5. Future MOD Labour Demand for Management Skills, 2004–2025 35

2.6.	Future MOD Labour Demands for Technical Skills, 2004–2025	36
2.7.	Future MOD Labour Demand for Structural Skills, 2004–2025	37
2.8.	Future MOD Labour Demand for Outfitting Skills, 2004–2025	38
2.9.	Future MOD Labour Demand for Support Skills, 2004–2025	39
2.10.	Scenario 1: Decreased MOD Requirements or Budgets—Labour Projections by Programme, 2004–2025	43
2.11.	Scenario 1: Decreased MOD Requirements or Budgets—Labour Projections by Skill Trade, 2004–2025	44
2.12.	Scenario 2: Addition of Future Submarine—Labour Projections by Programme, 2004–2025	46
2.13.	Scenario 2: Addition of Future Submarine—Labour Projections by Skill Trade, 2004–2025	47
2.14.	Scenario 3: Increased MOD Future Requirements—Labour Projections by Programme, 2004–2025	49
2.15.	Scenario 3: Increased MOD Future Requirements—Labour Projections by Skill Level, 2004–2025	49
2.16.	Level-Loading by Extending the Time Between Ship-Builds	52
2.17.	Impact of Moving Programmes to Avoid Peak Demand	53
2.18.	Impact of Both Extending and Moving Programmes to Avoid Peak Demand	54
2.19.	Level-Loading Labour Projections, by Programme	56
2.20.	Base Case and Level-Loaded Demands Compared with Current MOD Demand	56
2.21.	Level-Loading Labour Projections, by Skill Level	57
3.1.	UK Shipbuilding, Repair, and Offshore Employment in the 1990s	63
3.2.	Regional Shipbuilding, Repair, and Offshore Employment in the 1990s	64
3.3.	Share of Workers in Shipbuilding and Repair Subsectors in 2000	65
3.4.	Number of Workers in the Naval Yards, 1999–2003	65
3.5.	Age Profile of the Workforce in the Naval Shipyards in 2003	67

3.6.	Unemployment Levels in Important Shipyard Towns in 2002	72
3.7.	Number of Recruits in the Naval Shipyards, 1999–2003	74
3.8.	Total and Peak Outsourcing Undertaken by UK Naval Shipbuilding and Repair Companies	77
3.9.	Shipyard Labour Supply Model	80
3.10.	A Comparison of the Shipyard Labour Supply and Demand, 2004–2020	81
3.11.	Projected Shipyard Labour Supply by Skill Category Without Additional Recruitment, 2004–2020	82
3.12.	Forecast of the Naval Shipyard Management and Technical Workforce: Minimised Recruitment	83
3.13.	Forecast of the Naval Shipyard Manufacturing Workforce: Minimised Recruitment	84
3.14.	Forecast of the Shipbuilding Manufacturing Workforce: 8 Percent Recruitment Rate	85
4.1.	Ship Production Timeline	90
4.2.	Distribution of Final Assembly Facility Lengths	93
4.3.	Distribution of Final Assembly Facility Beams	95
4.4.	Distribution of Final Assembly Facility Draughts	95
4.5.	Distribution of Afloat Outfitting Facility Lengths	97
4.6.	Distribution of Afloat Outfitting Facility Beams	98
4.7.	Distribution of Afloat Outfitting Facility Draughts	99
4.8.	Type 45 Work Allocation	101
4.9.	Final Assembly Facilities Requirements for Type 45 Programme: VT Shipbuilding	103
4.10.	Final Assembly Facilities Requirements for Type 45 Programme: BAE Systems	104
4.11.	Afloat Outfitting Facilities Requirements for Type 45 Programme: BAE Systems	105
4.12.	Final Assembly and Afloat Outfitting Facilities Requirements for MOD Ships, 2004–2020	111
5.1.	Industrial Sectors of the Identified Suppliers	116
5.2.	Locations of Suppliers, by Country	117
5.3.	Suppliers' Ability to Take on More Work	119
5.4.	Long-Term Stability and Viability of Suppliers	120

5.5.	Dependence on Supplier/Competition	121
5.6.	Size of Suppliers, by Number of Employees in 2003	123
5.7.	Average Supplier Dependence on Different Sectors	125
5.8.	Percentage of Suppliers' Revenue Derived from MOD Ship Programmes	126
5.9.	Percentage of Suppliers' Revenue Derived from All Ship and Offshore Work	126
5.10.	Percentage of Suppliers' Revenue Derived from Military Work	127
5.11.	Numbers of Marine and Non-Marine Customers	128
5.12.	Numbers of Marine and Non-Marine Competitors	129
5.13.	Ease of Recruiting Four Classes of Employees	130
6.1.	Decline of Commercial Shipbuilding in the United Kingdom Over the Past Two Decades	139
6.2.	UK Offshore New-Build Market	139
A.1.	Schedule-Slip Scenario Labour Projections by Programme	166
A.2.	Schedule-Slip Scenario Labour Projections by Skill Level	168
A.3.	Base Case, Level-Loaded, and Schedule Slip Demands Compared to Current MOD Demand	169

Tables

S.1.	Future MOD Ship Programmes	xix
1.1.	Average Age and Full Ship Displacement of the Fleet	11
2.1.	Current MOD Shipbuilding Programmes	29
2.2.	Projected MOD Shipbuilding Programmes	29
2.3.	Scenario 1: Decreased MOD Requirements or Budgets Programme Assumptions	42
2.4.	Scenario 2: Addition of Future Submarine Programme Assumptions	45
2.5.	Scenario 3: Increased MOD Future Requirements—Programme Assumptions	48
4.1.	Number of Final Assembly Facilities at Naval Shipbuilders That Can Accommodate a Ship with a Given Beam and Length	96
4.2.	Number of Final Assembly Facilities at Naval Repair and Other Firms That Can Accommodate a Ship with a Given Beam and Length	97
4.3.	Number of Afloat Outfitting Facilities at Naval Shipbuilders That Can Accommodate a Vessel with Certain Length and Beam Characteristics	99
4.4.	Number of Afloat Outfitting Facilities at Naval Repair and Other Firms That Can Accommodate a Vessel with Certain Length and Beam Characteristics	100
4.5.	Ships Replaced by MARS	108
6.1.	Labour Resources of Medium-Sized Shipbuilders and Other Firms	142

7.1. Percentage Growth at Peak Demand Relative to Current
     Employment Levels 148
B.1. Basic Ship Dimensions 171
C.1. Skill Breakout, by Management/Technical and Manufacturing
     Categories 174

Summary

The United Kingdom's Ministry of Defence (MOD) is in the early stages of an ambitious effort to renew and upgrade its naval fleet over the next two decades through the production of new ships and submarines. Defence policymakers are seeking to gain a fuller understanding of the ability that shipyards, workers, and suppliers in the United Kingdom have to produce and deliver these vessels at the pace and in the order planned by the MOD.

This analysis, done at the request of the MOD's Defence Procurement Agency (DPA), focused on answering three fundamental questions: Can the existing shipbuilding industrial base meet future demands? Do problems exist with the numbers and types of facilities or the numbers and skills of the workforce? and If problems exist or can be anticipated, what can be done to alleviate them?

Relying both on public and proprietary data and on surveys of government and industry representatives,[2] the analysis addressed these questions by examining the capacity of the UK industrial base's current workforce and facilities, identifying the demands for these resources over the next two decades and exploring options to address situations in which future demands might exceed capabilities. The study aimed to help MOD policymakers (1) gain an understanding of the capacity of the United Kingdom's naval shipbuilding industrial base to successfully implement the MOD's current acquisition plan,

---

[2] By industry, we include naval shipbuilders and repairers; suppliers; design firms; and firms involved in commercial maritime work (i.e., offshore industry, commercial repairers, and producers).

and (2) gauge how alternative acquisition requirements, programmes, and schedules might affect the capacity of that industrial base.

MOD Ship Programmes[3]

The MOD is planning an extensive shipbuilding programme for the next 15 years, which can be divided into two main categories. The first category comprises programmes on contract that have already passed through the MOD's final approval process (Main Gate) and are somewhere in the demonstration and manufacture stage. The second category comprises prospective programmes that have yet to pass Main Gate but which the MOD anticipates will be built. Of course, the future procurement programmes continue to evolve in line with the strategic environment, financial imperatives, industrial developments, and new opportunities. Any statement of the programmes themselves, the number of ships, and the timings are speculative—particularly for the second category of programmes not yet past Main Gate. Thus, the reader must keep this caveat in mind when interpreting the results.

Table S.1 describes the *potential* future ship programmes and the potential size of their production runs.

Figure S.1 lays out the potential design and production timelines for the future programmes as identified in Table S.1.

The blue bars represent the programmes (or portions of programmes) that are past Main Gate and on contract. The grey bars represent the programmes that are either pre–Main Gate or potential additional procurements for the class that have not been contracted. The timings are a synthesis of our assumptions, data provided by the Integrated Project Teams (IPTs), and data provided by the shipyards. The dates are representative only and are not fixed as certain or are specific requirements from the Equipment Capability Customers.

---

[3] The information for this section is extensively drawn from the DPA's Web site (www.mod.uk/dpa/ipt/index.html) on the agency's current projects, the Royal Navy Web site (www.royal-navy.mod.uk), and Royal Navy (2003).

Table S.1
Future MOD Ship Programmes

Programme	Description	Potential Production Run
On Contract/Past Main Gate		
Astute-class attack submarine	The *Astute* is a new nuclear attack submarine (SSN) intended to replace the existing *Trafalgar* and *Swiftsure* classes. It is being designed for the support of the *Vanguard* ballistic nuclear submarine (SSBN), antisubmarine warfare, anti-surface warfare, surveillance and intelligence gathering, and land attack. There are three ships currently on order with the potential of six more acquired, for a total of nine.	9
Bay-class landing ship dock (LSD[A])	These new vessels are part of the Royal Fleet Auxiliary's (RFA's) rapid deployment force and will replace various landing ship logistic (LSL) classes in the RFA fleet. The LSD(A)'s main role is logistic support, bringing troops, trucks, tanks, and cargo into battle. It can also be used for humanitarian missions.	4
Type 45	Type 45 will be a multi-role destroyer whose principal mission is antiair defence (DDG). The first six ships of the class are currently on order. There is a potential for up to six additional ships to be procured. Both BAE Systems (Clyde Shipyards) and VT Shipbuilding[a] (Portsmouth) are involved in the production of these ships.	12
Prospective/Pre–Main Gate		
Future Aircraft Carrier (CVF)	CVF is the Royal Navy's next generation of aircraft carrier, meant to replace the current *Invincible*-class (CVS) carrier.	2
Future Surface Combatant (FSC)	FSC is notionally thought to be frigate-sized vessel and will replace the Type 23's and Type 22's currently in the fleet.	14
Joint Casualty Treatment Ship (JCTS)	JCTS is single-ship programme that will provide advanced medical capabilities to all three UK services. The ship can be used for combat operation support as well as humanitarian missions. As a 'grey hull' and therefore designated to operate within a task force, the ship will not be subject to the Geneva Conventions as would a conventional hospital ship.	1
Military Afloat Reach and Sustainability (MARS)	The MARS programme will be a series of ships (number and types currently undefined) that will provide supplies to the fleet and forces ashore. These supplies are a combination of dry goods and provisions, general stores, water, ammunition, and fuel products.	10

Table S.1—Continued

Programme	Description	Potential Production Run
Offshore Patrol Vessels– (OPV[H])	The Royal Navy has provisional plans to replace the *Castle* class of offshore patrol vessels now in use in the Falkland Islands. These new ships will have the ability to operate helicopters and will be leased on a similar basis as used with *River*-class patrol vessels.	2
Future Submarine	This submarine will be a follow-on to the *Astute* class. Its current size and mission are not yet defined.	7
Future Minehunter	This class will replace the minehunters currently in the Royal Navy fleet.	4

[a] Part of VT Group, formerly known as Vosper Thornycroft.

Figure S.1
Schedule of MOD Naval Programmes, 2005–2020

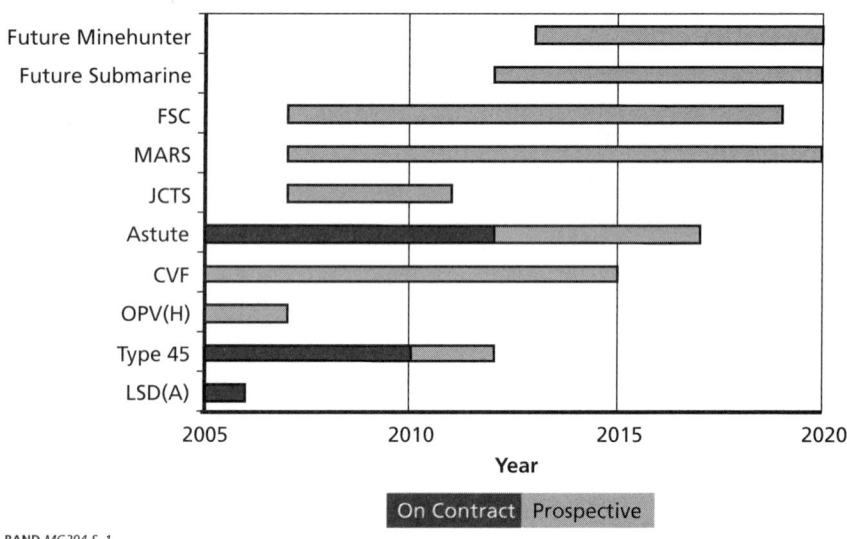

Figure S.1 shows that there will be periods when up to *six* programmes will be in various stages of design and construction. Not only will there be high concurrency, but these ships produced will be the largest of their type built in quite some time. This situation contrasts the current case in which there are only three programmes on contract. In 2010, for example, the Type 45, CVF, Astute, JCTS, MARS, and FSC programmes will all be under way simultaneously; however, not all these programmes will be in production at the same time. Nonetheless, the period between 2007 and 2013 is much busier for naval shipbuilding than has been seen recently.

Today, only a handful of UK shipbuilders will likely be able to produce these naval ships. After decades of consolidations and bankruptcies in the UK shipbuilding industry, only three major firms are currently involved in building ships for the MOD: BAE Systems, Swan Hunter, and VT Shipbuilding. In addition, there are three firms primarily involved in the repair of warships: Babcock Engineering Services, Devonport Management Limited, and Fleet Support Limited. For purposes of this report, the shipyards owned by these six firms are collectively termed 'naval shipyards'. An additional firm, Ferguson Shipbuilders, is also active but focuses mainly on the coastal patrol, ferries, fishery protection, and other commercial markets.

Policy Issues Pursued by RAND

This substantial MOD building programme, combined with the United Kingdom's diminished industrial base, raises a number of questions for defence policymakers:[4]

- Is the MOD shipbuilding plan feasible given the constraints of the industrial base?
- What is the programme's effect on the shipbuilders and ship repairers?

---

[4] An equally important issue, but beyond the scope of this study, is whether government can afford the shipbuilding plan. The increased level of shipbuilding activity will result in greater defence spending for naval acquisition. Whether this greater level of spending can be accommodated within the broader defence budget is not clear.

- Is the supplier base robust enough to meet the demand?
- Are there alternative timings for programmes that make the plan more robust?
- What is the effect if procurement quantities change?

At the request of the DPA, researchers of the RAND Corporation began addressing these questions in the autumn of 2003. Their main goal was to help MOD decisionmakers understand the capacity of the UK naval shipbuilding industrial base and its ability to undertake the MOD's shipbuilding programme over the next 15 years.

Study Structure

To analyse the issues facing the MOD, we decomposed the capacity evaluation into a supply and demand assessment in three distinct areas: labour, facilities, and suppliers. The study team evaluated these areas with respect both to the MOD's 'current plan' (which assumes that everything will be built as envisioned by the MOD's programme managers) and to several alternative shipbuilding scenarios:

- a pessimistic funding scenario, in which funding and/or requirements decrease such that fewer vessels are purchased
- an optimistic funding scenario, in which requirements and funding increase
- a new submarine scenario, in which a new, large submarine is designed and built
- a 'level-loaded' scenario, in which design and production timings are lengthened.

Our evaluations depended on two surveys that we conducted with firms involved in the shipbuilding industry and on interviews with government officials and industry associations. The first survey that the RAND team sent out requested information from several dozen firms involved in maritime design, repair, and production on their employment, future workload demands, facilities available, and their key suppliers. The second survey went to some 200 key suppliers that the firms had identified in the first survey. This latter survey asked the suppliers about their employment, the relative competitiveness of their market, their dependence on MOD and maritime work, and the challenges they anticipate in the future. After receiving both surveys, the RAND team held follow-on conversations with industry representatives to clarify data and discuss issues about which the surveys had not inquired.

In addition, the RAND team interviewed officials at a number of industry associations and government agencies.

How Will the MOD Programme Affect Shipyard Labour?

Labour Demand

This part of the analysis depended on a labour projection model that the RAND team developed.[5] The team used data obtained in the first survey of shipyards to populate this forecasting model, which allowed it to estimate future demands for labour emerging from the current MOD acquisition plan and from the alternative scenarios described above.

The team's analysis of the current MOD plan found that overlap of four large programmes—the Type 45, CVF, MARS, and Astute—is likely to cause demand for direct labour (all skills—e.g., management, technical, and manufacturing) to peak in 2009 at a level about 50 percent higher than the 2004 demand levels. Once past the peak, overall workload demand steadily declines for the foreseeable future. We show this workload projection in Figure S.2.

Structural and outfitting trades likely will show the most significant increase (in absolute terms). The technical workforce demand presents a more difficult challenge. The RAND team found that there

---

[5] See Arena, Schank, and Abbott (2004) for more details.

Figure S.2
Future MOD Labour Demands, by Programme

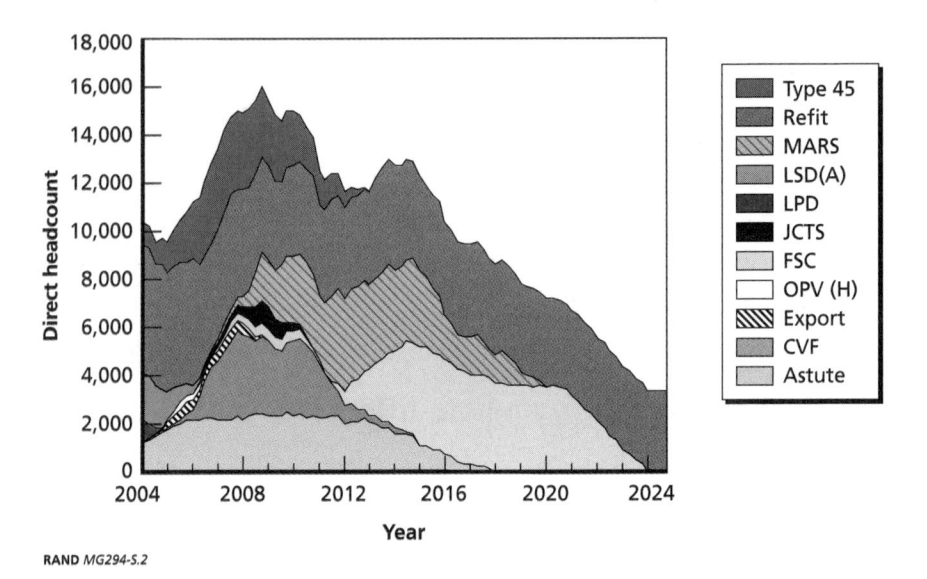

[Stacked area chart showing Direct headcount (0 to 18,000) from 2004 to 2024. Legend categories: Type 45, Refit, MARS, LSD(A), LPD, JCTS, FSC, OPV (H), Export, CVF, Astute.]

RAND MG294-S.2

could be a sharp drop-off in demand for the technical workforce[6] in the next two to three years, resulting largely from the rundown of the design work for the Type 45 and Astute. Thereafter, the trend reverses dramatically as CVF, MARS, and JCTS place near-simultaneous demand for technical workers. In the span of a few years, the demand for technical workers nearly doubles from its low.

With one exception, the other scenarios we explored also involved similar sharp increases in demand for production labour followed by steady decreases. Such increases in labour demand will force the naval shipbuilding and repair industrial base to rapidly increase its workforce, especially in specific outfitting, structural, and technical skills. Demands for technical workers under the alternative scenarios are much the same as the current plan. After an initial decline, the demand for these workers increases drastically. One

---

[6] Technical workforce includes the follow functions: design, drafting/CAD, engineering, estimating, planning, and programme control. See Appendix C for more detail.

notable exception was the scenario that involved a Future Submarine, which could make substantial demands for technical workers past 2010.

One way that the MOD could reduce these peak demands would be to level-load the ship production plan, which would involve starting programmes earlier or later, extending their build schedules, and increasing their build intervals. However, the MOD will need to consider operational needs in determining whether such an approach is feasible. Figure S.3 shows the change in total, direct employment for the current plan and a level-loaded example relative to the demand in 2004.

Labour Supply

The naval shipbuilding and repair industry will be challenged to meet the peak workforce demands as outlined above. To determine the nature and extent of that challenge, the RAND team built a spreadsheet model to forecast the labour supply for naval shipbuilders and repairers. The team then compared the potential supply picture with the projected total demand under the current MOD plan and the

Figure S.3
Direct Labour Demands for the Current Plan and a Level-Loaded Example

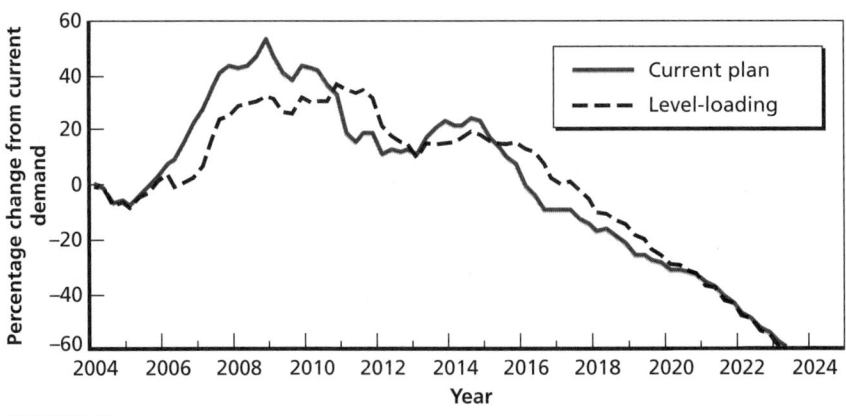

RAND MG294-S.3

level-loaded scenario described above. For the purposes of this analysis, the team combined the five skill subcategories into two: management/technical and manufacturing (structure, outfitting, and direct support). This simplification was required because of the limitations with data available.

As Figure S.4 indicates, the number of workers would drop from more than 12,000 workers to around 4,600 in the next 17 years if no steps are taken to replenish the workforce through hiring apprentices or experienced workers from other industries or from the ranks of the unemployed. We in no way suggest that the shipyards will not replace workers. In fact, several have active apprentice and recruiting programmes. Figure S.4 also shows how rapidly the current employment ranks decline because of the ageing of the workforce.

Figure S.4
Projected Shipyard Labour Supply Without Additional Recruitment, 2004–2020

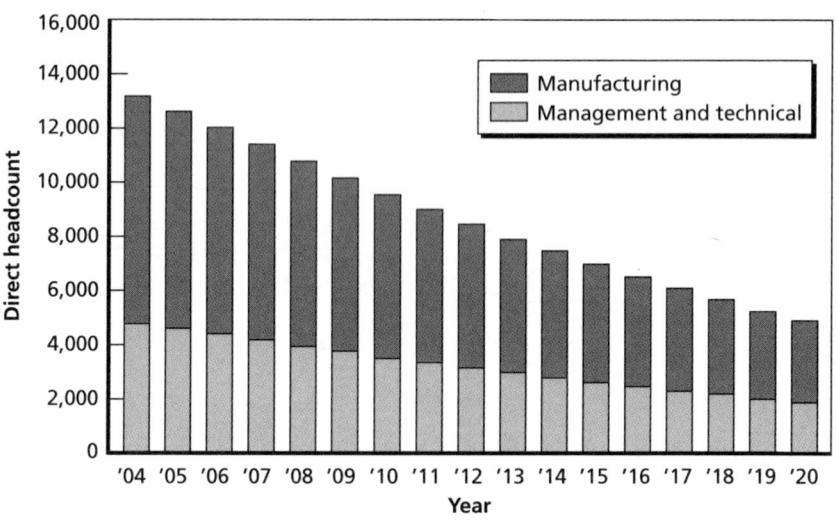

RAND MG294-S.4

Although there is no current shortage of workers, the shipyards expressed concern about their future ability to recruit particular skills (e.g., design, electrical, test and commissioning, and steel work). Many of the shipyards have begun apprentice programmes in recognition of the ageing problem that are aimed at maintaining current or core workforce levels and not necessarily to meet future peak workload.

There are, of course, other labour sources from which the shipyards can draw workers. For example, some of the shipyards have recently made workers redundant and may be able to rehire these former workers. There is also the opportunity to draw workers from related industries and from among the general unemployed. Another alternative for the shipyards is to rely more heavily on outsourced activities, a trend that has been increasing as of late.[7]

Despite these additional sources of labour, the RAND team concluded that it will be difficult for the shipyards to grow to meet peak labour demands. Figure S.5 shows workload demand for the baseline and level-loading scenarios along with employment projections. Assuming a modest growth rate, the shipyards as a whole may not be able to meet peak labour demand for production workers. Even under the level-loaded scenario, shipyards will approximately meet the peak production demands. For the technical workforce, there are currently enough workers at the firms to grow to the needed peak levels, but only if these workers are retained through the near-term downturn. In all, meeting the peaks in workload demand will require that shipyards share work to a greater extent than they do now.

These supply and demand results present labour issues at an aggregate (macro) level to simplify the presentation and to protect business-sensitive information. The macro results are useful in that they portray the magnitude and the timing of the labour issues the

---

[7] Of course, there is a limit to the extent that these activities can be outsourced. See Schank et al. (forthcoming) for more detail.

Figure S.5
Shipyard Employment Projections Versus Demand
(baseline and level-loaded)

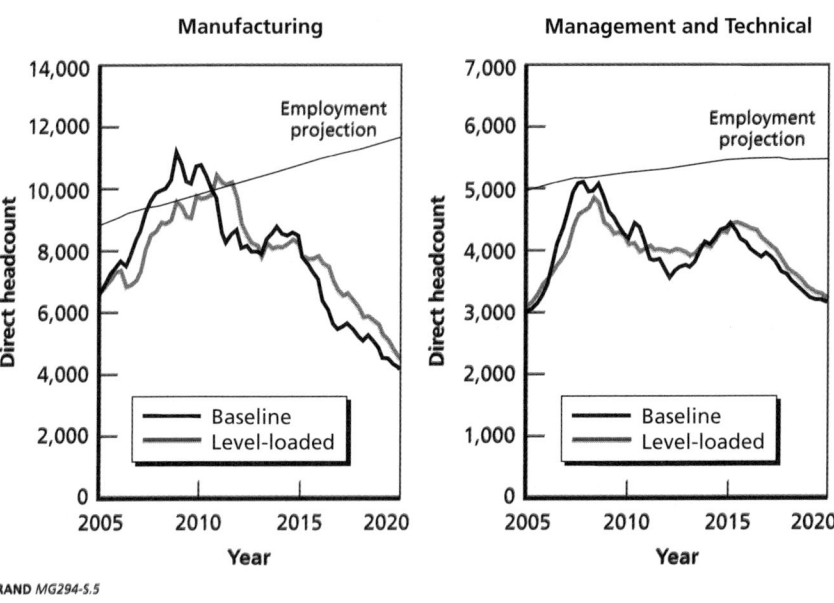

industry faces. However, the macro view masks the effect at an individual shipyard or firm level. In a sense, the macro view could be interpreted as a case in which there is an unlimited ability to share work between firms. Thus, where the macro-level demand might appear reasonable or achievable, there may be problems of either irregular or high demand for labour at an individual firm making such a plan difficult to implement.

How Will the MOD Programme Affect Shipyard Facilities?

In this part of the analysis, the RAND study team focused on the facility implications of the current MOD plan. In particular, the team concentrated on final assembly facilities (docks, slipways, land-level areas, etc.) and afloat outfitting locations (mainly piers and quays).

The facilities considered included those at the naval shipyards and some commercial yards.

Not surprisingly, the RAND team found that different programmes likely will stress different facilities. In general, as Figure S.6 indicates, the demand for final assembly facilities will be particularly high between 2006 and 2010.[8]

The Type 45 programme will create a substantial demand for final assembly and outfitting locations because of the build interval between ships (assumed to be six months). Although the facilities on the Clyde might be able to handle this schedule (with facilities upgrades and some careful scheduling), extending the build interval of the Type 45 to nine months might help to alleviate any potential problems and make the build schedule more robust.

Figure S.6
Total UK Final Assembly and Afloat Outfitting Facility Requirements, 2004–2020

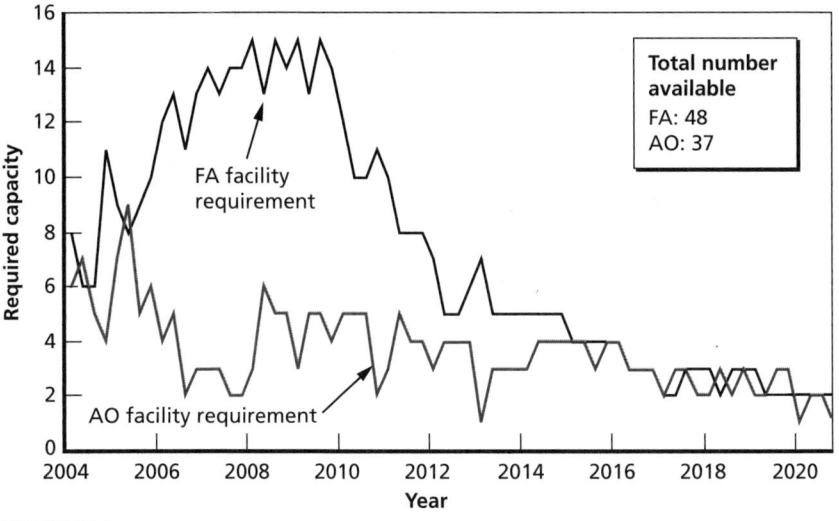

RAND MG294-S.6

---

[8] Because of the sensitivity of the data, this demand does not include all refit and repair work.

The sheer size of the ships in the CVF and MARS programmes poses challenges. CVF assembly will require some facilities upgrades and investments because no assembly location today could handle the CVF ships without modification or upgrade. Further complicating the picture is whether the final CVF assembly location will also be used to build large block portions of the ship. There is a potential overlap between the assembly of the first hull and the production of blocks for the second hull. This overlap implies that the second hull's blocks will either need to start construction outside the final assembly dock or be delayed until the first hull leaves the dock.

Similarly, large MARS ships will be equally challenged for a final assembly location. Although there are facilities in the United Kingdom that could construct these ships, it is likely that at least two facilities (or a facility that can construct two ships at once) would be needed based on the notional delivery schedule of one ship per year. In most cases, any of these candidate facilities would need to be upgraded or reopened—thus requiring investment.

How Will the MOD Programme Affect Shipbuilding Suppliers?

More than half the unit cost value of a naval vessel is provided by firms other than the shipbuilder.[9] So the ability of suppliers to meet the demand based on the MOD's plans is an important consideration in addressing the UK industry's capacity. The study's surveys of both the shipyards and the suppliers indicate that there will be generally few issues surrounding the increased workload for the suppliers. For the most part, the suppliers do not rely on MOD business, so they are less subject to the variations in demand (in contrast with the shipyards). Further, most of the suppliers are based in the United Kingdom. However, these suppliers have indicated that the uncertainty in

---

[9] *The Clyde Shipyards Task Force Report* (2002, p. 41).

the MOD's programme hinders their ability to plan and invest in a timely way.

The Role of Nontraditional Sources for Naval Shipbuilding

The UK commercial shipbuilding and offshore industry have resources that may help to produce ships for the MOD. These resources are currently underemployed because of downturns in both sectors; therefore, these resources could be available to the MOD for its shipbuilding programme. Medium-sized shipbuilders[10] have had a role in the past and could play a role in the production of future ship classes—from a builder of blocks and modules to a producer of smaller vessels. The offshore industry also has the potential to contribute to the programme. These firms have facilities and labour resources that could be employed—most notably in management and technical skills. However, its role, if any, will need to be carefully matched to its capabilities and skills. For programmes that are similar to commercial products, the commercial/offshore industry could play a broader role in management, design, and production. However, for combatant ships, their role will be more limited.

Issues for the MOD to Consider

The RAND team made short- and long-term observations for the MOD.
　In the short term, the MOD could:

- *Consider ways to level-load the labour demand.* In essence, the MOD will need to carefully consider the timings of various programmes. Some programmes will need to be shifted later, while others may need to have increased build intervals. It may also be

---

[10] Appledore (owned by DML), Ferguson Shipbuilders, and Harland and Wolff.

necessary to shift design work earlier to mitigate the near-term lack of demand (the near-term dip in technical labour demand). The DPA will need to consider the labour effects at the individual firm or shipyard level to achieve the benefits of level-loading. Thus, any level-loading plan will need to be developed in consultation with industry.

- *Work with the Department of Trade and Industry to encourage training in skills that are in demand outside the shipbuilding and repair industry.* The UK government and shipbuilding industry should focus on training skills that are readily employable outside shipbuilding. In this way, any resulting unemployment can be minimised.
- *Consider relaxing the shipbuilding industrial policy to mitigate problems resulting from peak demand.* The MOD should re-examine industrial policy with respect to obtaining work content overseas. For example, the policy might allow UK shipyards to obtain major units or subassemblies from abroad in cases in which there is peak demand and it is not possible to easily obtain that content domestically.
- *Encourage the use of more outsourcing.* One way that commercial shipbuilders manage variable workloads is to employ outsourcing vendors that provide services and goods.
- *Evaluate the future of shipbuilding at Barrow.* With the current realignment within BAE Systems, the Barrow-in-Furness facility is exclusively dedicated to submarine production. The end of surface ship building in Barrow resulted in significant redundancies and the closure of some facilities. Barrow remains an untapped source of production capability and could likely play a significant role in the coming shipbuilding programme.[11]
- *Consider the use of medium-sized shipyards to meet some of this demand peak.* Shipyards such as Ferguson, Appledore, and Harland and Wolff could play a role in meeting the peak demands by constructing blocks or structural units. Ferguson and Apple-

---

[11] Since the original writing of the report, BAE Systems Naval Ships has stated that it is now possible to use Barrow surface ship capability.

dore could produce smaller naval vessels, like the survey vessels both have produced in the last few years. Harland and Wolff was, at one time, capable of producing large auxiliary vessels. Whether that capability could be cost-effectively reestablished remains to be seen.
- *Explore the utilisation of facilities for Type 45, CVF, and MARS.* There may be facility challenges for these programmes, and the MOD needs to understand where there are potential conflicts and the actions that can be taken to mitigate them.
- *Have the Supplier Relations Group (SRG) investigate suppliers that are thought to be at risk.* In our surveys, the shipyards identified certain suppliers they felt were at risk. It might be worthwhile for the SRG to interact with the shipyards and suppliers to better understand the ones at risk and any corrective actions required.

In the long term, the MOD should, among other things:

- *Make long-term industrial planning part of the acquisition process.* This type of planning must become part of the process that the MOD uses to define the timing for the various naval requirements. The potential benefits of long-term planning include the ability to understand financial implications, reduce cost and schedule risks, and anticipate future problems. A strategic examination of the overall build programme with respect to the industrial impact should be done at least annually with an outlook of 10 to 15 years.
- *Define an appropriate role for the offshore industry.*[12] Better work-sharing between the shipyards will be necessary to meet the peak labour demand. The offshore industry may help the naval shipbuilding industry bring this collaboration about. Although the offshore industry might not feature strongly in direct fabrica-

---

[12] That is, those firms involved in the design, manufacture, and support of capital facilities for oil and gas in the sea (mainly the North Sea for the United Kingdom).

tion, it might feature more prominently in assembly and integration.
- *Carefully consider the implications of foreign procurement of complete ships.* Because foreign procurement carries risks, we recommend that the MOD thoroughly take into account issues such as access to technology and political disruptions before procuring entire vessels from abroad. The UK government could allow shipyards to consider outsourcing work to foreign sources when there is a need to reduce a labour peak, workers are not available elsewhere in the United Kingdom, and workers would only be needed for a short period.
- *Encourage long-term investment through multi-ship contracts.* Most naval shipyards have not modernised facilities during the past several years. Longer-term contracts will allow the shipyards to justify this type of major investment. However, the experience of US programmes has shown that multi-ship contracts work best for mature designs. So, such an approach may not include the first-of-class ship.
- *Consider the feasibility of competition in light of industrial base constraints.* Competition may not always yield better prices or result in a balanced allocation of work under conditions in which there are high resource demands. In such an environment, it is possible that there will be fewer potential bidders on subsequent programmes, that bidders will take on more work than is optimal, or that shipyards will be less inclined to cooperate for fear of losing a competitive advantage.
- *Explore the advantages of common and/or compatible three-dimensional computer-assisted design/computer-aided manufacturing (CAD/CAM) design tools.* The MOD might want to facilitate a discussion among shipbuilding firms (and potentially include the CAD/CAM vendors) to explore whether they should adopt a common or interoperable design tool or establish standards so that design work can be easily interchanged.

Acknowledgements

This report would not have been possible without the contributions of many firms and individuals. First, we would like to thank Andy McClelland, Andrew Stafford, and David Twitchin of the DPA for guiding and helping to coordinate the study within the agency. We would also like to thank Stephen Highfield, also of the DPA, for his assistance and coordinating our interaction with the Pricing and Forecast Group. The authors would like to acknowledge the shipbuilding IPT leaders and members who participated in this study and provided insight to and data for their programmes. Our gratitude goes to Stephen French, Director General—Equipment, and the marine Directors of Equipment Capability (DECs) for their views on the future shipbuilding programme. Also of the DPA, we are grateful to Rear Admiral Ric Cheadle for his insightful comments on an earlier version of this report.

At RAND, we would like to thank the reviewers, Robert Murphy and Irv Blickstein, for the many improvements and suggestions they made to the text. The authors would also like to acknowledge the assistance of Nathan Tranquilli and Ricardo Basurto-Dávila for their help with data analysis and modelling. We also thank Joan Myers for coordinating the document preparation.

Finally, we are deeply indebted to the shipbuilders, marine firms, contractors, suppliers, and associations that participated in the study survey and interviews. Without their assistance and cooperation, this research would not have been possible.

Abbreviations

AO	afloat outfitting
BES	Babcock Engineering Services
BMT	British Maritime Technologies
CAD/CAM	computer-assisted design/computer-aided manufacturing
CVF	Future Aircraft Carrier
CVS	*Invincible*-class carrier
DDG	antiair defence
DDH	Devonshire Dock Hall
DEC	Director of Equipment Capability
DML	Devonport Management Limited
DPA	Defence Procurement Agency
DTI	Department of Trade and Industry
ECC	Equipment Capability Customer
FA	final assembly
FSC	Future Surface Combatant
FSL	Fleet Support Limited
HVAC	heating, ventilation, and air conditioning
IPT	Integrated Project Team
JCA	Joint Combat Aircraft
JCTS	Joint Casualty Treatment Ship

KBR	Kellogg-Brown and Root
LPD(R)	landing platform dock (replacement)
LPH	helicopter landing platform
LSD(A)	Landing Ship Dock Auxiliary
MARS	Military Afloat Reach and Sustainability
MOD	Ministry of Defence
ONS	Office of National Statistics
OPV	offshore patrol vessel
OPV(H)	Offshore Patrol Vessel–Helicopter
PFG	Pricing and Forecast Group
RFA	Royal Fleet Auxiliary
SEMTA	Sector Skills Council for Science, Engineering, and Manufacturing Technology
SRG	Supplier Relations Group
SSN	nuclear attack submarine
STOVL	short take-off, vertical landing
VAR	voluntary attrition rate
VT	Vosper Thornycroft

CHAPTER ONE
Introduction

Between September 2003 and April 2004, RAND researchers analysed the capability of the shipbuilding industrial base in the United Kingdom to meet the demands of current and future Ministry of Defence (MOD) programmes. The MOD is in the early stages of an ambitious effort to procure upwards of 50 naval ships and submarines over the next two decades, and defence policymakers are seeking to gain a fuller understanding of the ability of UK shipyards, workers, and suppliers to produce and deliver those vessels at the pace and in the order planned by the MOD.

This analysis, done at the request of the MOD's Defence Procurement Agency (DPA), focused on answering three fundamental questions: Can the existing shipbuilding industrial base expand to meet future demands? Do problems exist with the numbers and types of facilities or the numbers and skills of the workforce? and If future problems do exist, what can be done to alleviate them?

Relying both on public and proprietary data and on surveys of government and industry representatives, the analysis addressed these questions by examining the capacity of the UK industrial base's current workforce and facilities, identifying demands for those resources during the next two decades, and exploring options to address situations in which future demands might exceed available capabilities. The study aimed to help MOD policymakers (1) gain an understanding of the capacity of the United Kingdom's naval shipbuilding industrial base to successfully implement the MOD's current acquisition plan, and (2) gauge how alternative acquisition requirements, programmes, and schedules might affect the capacity of that industrial base.

Warship Production Is a Unique Industry

The design and construction of warships[1] is one of the more complicated weapon system engineering and manufacturing tasks that a country can undertake. Warships require a complex integration of communication, control, weapons, and sensors that must work together as a coherent system. These components, or subsystems, are a mix of various technologies (e.g., electronics, mechanical systems, software). Oftentimes these technologies (particularly weapon systems) are state of the art or are undergoing development at the time a programme begins.

Moreover, warships must generate power to operate these systems and propel the ship. For surface ships, the method of power generation is typically gas turbines or diesel engines. Submarines generate power from diesel engines or nuclear reactors. Similarly, aircraft carriers can be conventionally or nuclear powered.

Warships must also be able to survive attack and protect its crew onboard. This protection ranges from the use of armour to systems designed to isolate the crew from chemical, biological, and radiological attack. Much of the equipment is shock hardened or isolated from blast shock. Some vessels also act as platforms for air vehicles, such as aircraft or helicopters.

Beyond their direct military mission capability, warships must perform another set of functions: housing and feeding the hundreds of sailors ('hotel' functions). Warships also need to provide for the health of the crew and thus require medical facilities. All these capabilities must be sustained for up to several months, requiring a sig-

---

[1] For the purpose of discussion here, *warship* refers to a class of ship or submarine that is blue-water capable. That is, the ship can fulfil a role beyond coastal protection. Also, we include ships beyond the direct combatants, such as logistics ships.

nificant amount of equipment and provision storage. These non-mission capabilities make warships unique compared with other military assets, such as tanks and aircraft.

To house all these functions and capabilities, warships must be *large* weapon systems. For example, the crew size for a warship can number in the hundreds, and in some case thousands. These ships weigh thousands of tonnes (displacement) and can reach lengths of a few hundred metres.

Given the size and complexity, manufacturing warships requires substantial design, engineering, management, testing, and production resources. For even a modest vessel, the engineering and design workforces can peak well in excess of 100 staff. Several engineering specialties are involved in shipbuilding (e.g., electrical, mechanical, naval architecture). The design of modern naval ships is now done using sophisticated three-dimensional computer-assisted design (CAD) tools. Thus, the design workforce must be highly skilled and educated. Production also requires many proficient skills or trades, such as electricians, welders, and painters. Testing these complex systems also requires commissioning and test specialists to verify functionality. For certain skills, it might take years to become proficient (e.g., nuclear-qualified welders and commissioning engineers). The workforce for the production trades might peak in the thousands for a typical naval vessel.

The manufacture of warships also requires significant facilities. Shops and plate lines/steel fabrication facilities make component parts and structure. Docks, slipways, piers, and cranes are used for assembly and integration activities. These facilities occupy large areas of land and must be water accessible (often the more valued real estate of any country). The facilities themselves are expensive to build and maintain. Hence, shipbuilding is both a capital- and labour-intensive industry. As such, it cannot be developed or expanded without significant resources, planning effort, and a long lead time.

Another aspect to naval production is that it relies on a significant vendor base; these vendors supply products ranging from services to material and equipment. Painting services, modular unit (such as accommodations) manufacturers, material suppliers, and sophisticated weapon systems providers (such as fire control and sonar systems) are examples of the diversity of the vendor base. Thus, the shipbuilding industrial base encompasses a much broader array of firms beyond just the shipyards.

Nations that maintain a significant navy (particularly one with expeditionary capabilities) have established domestic industries that are specialised in warship production.[2] This fact contrasts with the commercial shipbuilding industry, in which a small number of countries dominate the production in a particular market segment (such as cruise ships or bulk containers).[3] There are many reasons given as to why a domestic naval shipbuilding capability might be desirable.[4] Although not advocating for or against these reasons, we will briefly discuss the issue to illustrate the broader issues involved in naval acquisition.

One reason a domestic industry might be necessary is that certain technologies, such as nuclear propulsion, are sensitive and therefore difficult to obtain from outside sources. A second reason could be political, because most shipbuilding industries employ large numbers of workers. Thus, politicians have a vested interest in preserving a healthy domestic industry. But politicians are not alone in their interest in protecting large employers (such as shipyards). Shipyards in the United Kingdom are long-established firms with extensive histories, which makes them even more politically difficult to close. Also, warships are paid with public funds. It is argued that these public funds should be spent to support or stimulate the domestic economy. A further argument given is security. By maintaining a domestic source of production a country is better able to control information about the technologies and capabilities of its warships. A final reason given in favour of having a domestic industry is that it can tailor the warships better to meet a country's specific needs or operating doctrine. In buying vessels overseas, governments may find it more diffi-

---

[2] Todd and Lindberg (1996).

[3] Birkler et al. (forthcoming).

[4] Todd and Lindberg (1996).

cult to obtain customised ships or may be restricted to only standard designs.

Recent trends in foreign purchases of warships (from both the United Kingdom and other nations) suggest that design and possibly first-of-class are purchased abroad and follow-on ships are manufactured domestically.[5] For the United Kingdom, shipbuilding is also tied to national pride, since its history of the industry goes back centuries. The UK empire grew, in large part, because of a strong navy and merchant fleet. Whatever the reason for favouring domestic production, maintaining a strong navy seems to be inexorably tied to a domestic shipbuilding industry.[6]

MOD Ship Programmes[7]

Over the next 15 years, the MOD is planning an extensive shipbuilding programme. During this period, the MOD will be enhancing the capabilities of the fleet as well as renewing it. The MOD's shipbuilding programme can be divided into two main categories: those programmes that have already passed through the MOD's final approvals process (Main Gate) and are somewhere in the demonstration and manufacture stage, and those programmes that have yet to pass Main Gate but which the MOD anticipates will be built. Below, we will describe the programmes in both categories.

Of course, the future procurement programmes continue to evolve in line with the strategic environment, financial imperatives, industrial developments, and new opportunities. Any statement of

---

[5] For example, the *Super Vita* class for the Hellenic Navy, based on a VT design, is produced in Greece. See Birkler et al. (forthcoming) for more detail.

[6] There are, of course, global firms that do successfully export warships (e.g., some German manufacturers). BAE and VT export design and production warships to countries such as Malaysia, Brunei, Oman, Qatar, and Greece; however, it is a smaller portion of their overall work.

[7] The information for this section is extensively drawn from the DPA's Web site on the agency's current projects (http://www.mod.uk/dpa/ipt/index.html), the Royal Navy's Web site (http://www.royal-navy.mod.uk), and Royal Navy (2003).

the programmes themselves, the number of ships, and the timings are speculative—particularly for the second category of programmes not yet past Main Gate. Thus, the reader must keep this caveat in mind when interpreting results.

Programmes Past Main Gate
As of the writing of this report, the following three programmes were in the design and production stage.

Astute-Class Attack Submarine. The Astute programme is a new nuclear attack submarine (SSN) intended to replace the existing *Trafalgar* and *Swiftsure* classes. The tasks of the attack submarine are stated to be the following: support of the *Vanguard* class, antisubmarine warfare, anti-surface warfare, surveillance and intelligence gathering, and land attack. Currently, three boats are on contract with BAE Systems, which is producing the boats at its facilities in Barrow. Additional submarines of the class may be procured—potentially up to six; however, the MOD has made no decision as to the number beyond the initial three vessels. The *Astute* class uses nuclear propulsion and will displace approximately 7,800 metric tons (submerged). The crew complement is notionally 98 sailors. The length overall of the vessels is 97 metres with a beam of approximately 11 metres. The vessels will be armed with Tomahawk cruise missiles and Spearfish torpedoes.

Landing Platform Dock (Replacement)—LPD(R). The LPD(R) is a programme designated to replace the fleet's amphibious capability. These replacement ships will provide landing capabilities for up to eight landing craft, which carry various combat vehicles and forces. The ships will displace approximately 18,500 metric tons (full load) and have a crew complement of 349 and an additional berthing capability for up to 304 troops. The ships are also capable of having helicopter operations from the flight deck; up to two medium-sized helicopters can be accommodated. These ships will replace the fleet's two amphibious ships: HMS *Fearless* and HMS *Intrepid*. The first of the LPD(R) ships (HMS *Albion*) is already in service, and the second (HMS *Bulwark*) is expected to enter service in early 2005. The ships were built at BAE's facilities in Barrow. The length of the vessel is approximately 180 metres with a maximum beam of about 29 metres.

Bay-Class Landing Ship Dock—LSD(A). These new vessels are part of the Royal Fleet Auxiliary's (RFA's) rapid deployment force. These ships will replace the various landing ship logistic classes in the RFA fleet. The LSD(A)'s main role, as the designation implies, is logistic support by transport of troops, trucks, tanks, and cargo into battle. The ships can also be used for humanitarian missions. The design of the *Bay* class is a modification of the Dutch *Rotterdam* class. Swan Hunter and BAE Systems (Clyde Shipyards) are each producing two vessels of the class. The ships have an approximate length overall of 177 metres and a beam of 27 metres. The full displacement of the vessels is approximately 16,160 metric tons, and the crew size is approximately 60.

Type 45. The Type 45 is a multi-role destroyer whose principal mission is antiair defence (DDG). The first six ships of the class are currently on order, and there is a potential for up to six additional ships to be procured. BAE Systems Naval Ships (Clyde Shipyards) is the prime contractor for the Type 45. VT Shipbuilding (Portsmouth) is a subcontractor on the programme. Each firm produces specific blocks for all the ships in the class, which are subsequently assembled at BAE's Clyde Shipyards. The main weapon system on the ship, the Principal Anti-Air Missile System, is being jointly developed with the French and Italian navies. The ships could also be fitted with cruise missiles and a 155-millimetre gun, but currently there is no requirement for either system. The Type 45 ships are notionally replacing the current Type 42 class. With approximate length overall of 152 metres and a beam of 21 metres, the Type 45 displaces 7,350 metric tons. The crew complement is about 190.

Projects Pre–Main Gate[8]

Beyond the projects listed above, there are several programmes in the conceptual planning phase that have yet to pass through Main Gate

---

[8] Main Gate approval is the point at which DPA authorises a programme to proceed to the demonstration and manufacturing stage.

approval. As such, there is little that can be said definitively of either the ship characteristics of or the contractors involved in these programmes.

Future Aircraft Carrier (CVF). The CVF is the Royal Navy's next generation of aircraft carrier and is meant to replace the current *Invincible*-class (CVS) carriers. The ships will be considerably larger than previous UK aircraft carriers and potentially larger than any ship produced for the Royal Navy in decades. The size is a result of the desire to support an enhanced strike capability over the existing CVS class. Two CVF class ships will be produced. The base design of the CVF is configured for short take-off, vertical landing (STOVL) aircraft operation but can be adapted to conventional flight operations should such a requirement materialise or the Royal Navy employ different aircraft. These ships will carry a mix of F-35 aircraft (Joint Combat Aircraft [JCA] STOVL variant) and helicopters.

The MOD has announced that it intends to deliver the CVF through an alliance approach comprising BAE Systems, Thales UK, and the MOD. The contracting arrangements have yet to be finalised. BAE and Thales UK have formed the Aircraft Carrier Team to take the programme through the assessment phase leading up to Main Gate. Although the shipbuilding strategy continues to evolve, it is likely that multiple shipyards will be involved with CVF production, given its size and complexity. The planned in-service date for the first CVF is 2012. No crew size has been stated for the ship.

Future Surface Combatant (FSC). An Integrated Project Team (IPT) has recently been formed for the FSC programme. The ship is notionally thought to be frigate-sized vessel and will replace the Type 23s and Type 22s currently in the fleet. The FSC requirements are stated as a multi-role/adaptable ship capable of surface and submarine defence, shore support, homeland defence, marine interdiction, and shore forces deployment. Early concepts are exploring both monohull and trimaran hull forms.

Joint Casualty Treatment Ship (JCTS). The JCTS is a single ship programme that will provide advanced medical capabilities to all three UK services. The ship can be used for combat operation support as well as humanitarian missions. As a 'grey hull' and therefore designated to operate within a task force, the ship will not be subject to the Geneva Conventions as would a conventional hospital ship. The JCTS will represent a significant enhancement of the capabilities over the aviation training ship RFA *Argus*, which it replaces. According to the DPA's Web site, 'Subject to further study, the total JCTS requirement is expected to be 8 operating tables and not less than 150 beds (expressed as >150)'.[9] The JCTS will also have a flight deck capable of handling two helicopters. There is currently no in-service date stated for the ship.

Military Afloat Reach and Sustainability (MARS). The MARS programme is a series of ships (number and types currently undefined) that will provide supplies to the fleet and forces ashore. These supplies include a combination of dry goods and provisions, general stores, water, ammunition, and fuel products. The ships are thought to have a multi-commodity capability (i.e., the ability to carry more than one type of cargo). However, this capability is subject to change. The series of MARS ships will replace the RFA tankers, AORs (Auxiliary Oiler Replenishments), and AFSHs (Auxiliary Fleet Support Helicopters). Compared with existing ships, MARS ships will be double-hulled tankers, making them compliant with EU shipping regulations. The MARS programme will also provide a variant configuration for use as a sea base.

Offshore Patrol Vessel Helicopter (OPV[H]). The Royal Navy has provisional plans to replace the *Castle* class of offshore patrol vessels, now in use in the Falkland Islands, with OPV(H)s. These new ships will have the ability to operate helicopters and will be leased on a similar basis as used with *River*-class patrol vessels.

Other Speculative Programmes

Beyond the programmes currently being run by IPTs, two other programmes might be undertaken during the period we explore. The first is a new submarine (called Future Submarine) that the RAND team assumed to be of the size of a *Vanguard*-class submarine. The

---

[9] www.mod.uk/dpa/ipt/jcts.html (last accessed November 2004).

second would be a replacement of the minehunter capability. For this programme, the team assumed that the ship would be a modified FSC.

Notional Programme Timelines

Figure 1.1 displays an illustrative, high-level timeline in Gantt format for the design and production of the future programmes as described above. The blue bars represent the programme (or portions of programmes) that are past Main Gate and are on contract. The grey bars represent the programmes that are either pre–Main Gate or potential additional procurements for the class that have not been contracted. The timings are a synthesis of our assumptions, data provided by the IPTs, and data provided by the shipyards. The dates are representative only and are not fixed as certain or are specific requirements from Equipment Capability Customers (ECCs).

Figure 1.1
Gantt Chart of MOD Naval Programmes over the Next 15 Years

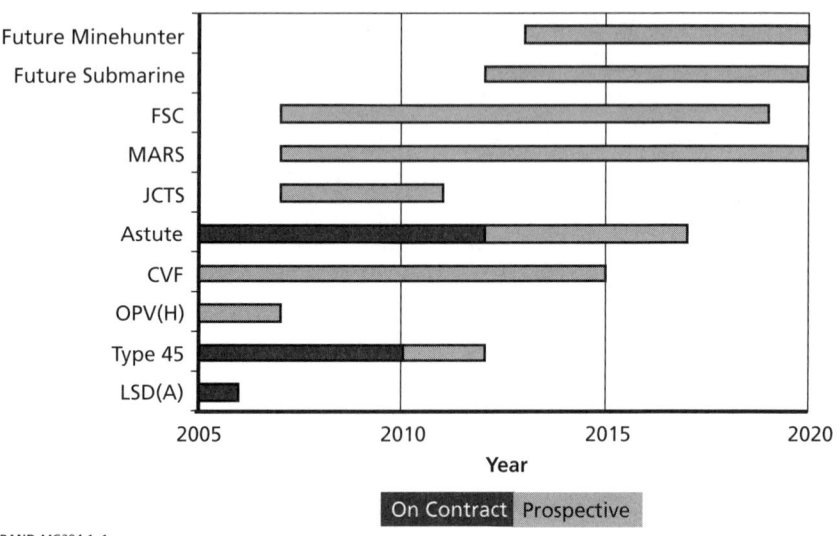

RAND MG294-1.1

It is worthwhile to note that there will be periods when there could be up to *six* programmes in various stages of design and construction. This level contrasts the three that are on contract as of early 2004. Not only will there be increased concurrency, but these ships produced will be the largest of their type built in quite some time. Take 2010 as an example. The Type 45, CVF, Astute, JCTS, MARS, and FSC programmes will all be under way at the same time. However, not all these programmes will be in production. The FSC, for example, will likely be in its design phase in 2010. However, the period between 2007 and 2013 is much busier for shipbuilding than has been seen recently.

The implication for such a build plan is that the MOD will radically transform its fleet by 2020. Table 1.1 shows an estimate for the average age and size of the fleet in 2004 and 2020. For the sake of consistency, the data exclude the Antarctic patrol, training, and repair vessels, inasmuch as it is uncertain whether the MOD will replace these vessels. The table shows that the average age drops from about 16 years to 12.5 years. Furthermore, the average displacement weight increases by more than 40 percent. Thus, the MOD will have younger (27 percent) and larger (48 percent) vessels in the fleet by 2020 than it does today.

Table 1.1
Average Age and Full Ship Displacement of the Fleet

Calendar Year	Average Age (years)	Average, Full Displacement (metric tons)
2004	15.9	9,200
2020	12.5	13,600

Business Environment of UK Naval Shipbuilding Industrial Base Since 1985

The ability of the UK shipbuilding industry to accommodate these future plans of the MOD is constrained by the history of naval shipbuilding of the past few decades. The end of the Cold War resulted in a profound reduction in naval shipbuilding for the United Kingdom as requirements lessened and the country sought to capitalise on the 'peace dividend'. Figure 1.2 shows how the Royal Navy combatant fleet has shrunk over the past three decades. There has been a steady decline in the fleet size since 1970: In 2000, the combatant fleet was about 60 percent of its 1970 size.

Figure 1.2
Royal Navy Fleet Size over the Past Four Decades

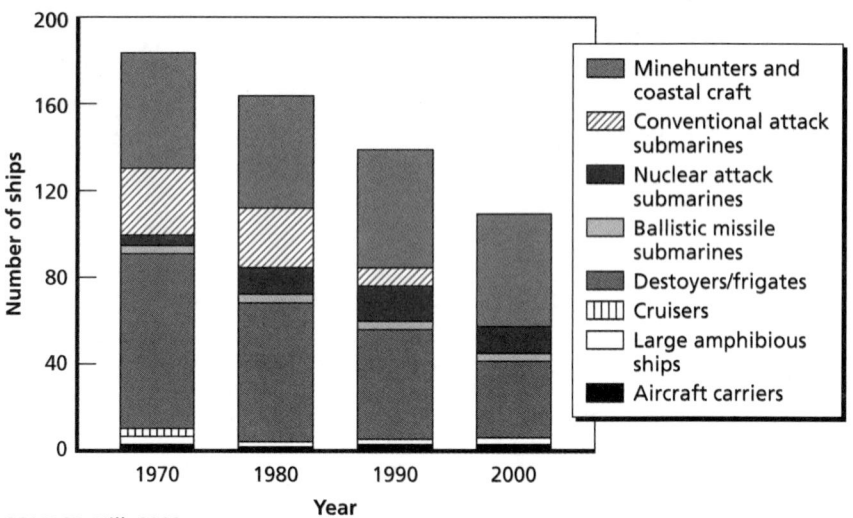

[Bar chart showing number of ships from 1970 to 2000, with categories: Minehunters and coastal craft, Conventional attack submarines, Nuclear attack submarines, Ballistic missile submarines, Destroyers/frigates, Cruisers, Large amphibious ships, Aircraft carriers]

SOURCE: Hill, 2002.
RAND MG294-1.2

This decline in the fleet size has also resulted in fewer new orders. In 1970, the fleet size of the combatant force was 184 ships. If the average ship lasts 30 years, then about six new ships would have been needed to be brought into the fleet each year to maintain the total fleet size. This value, of course, assumes that the fleet age is uniformly distributed. In 2000, the corresponding annual replacement rate would be about 3.5 ships per year. In fact, these crude approximations correspond quite well to the actual ship delivery history over this period. Figure 1.3 shows the annual delivery of combatants between 1970 and 2000. In the 1970s, the number of ships delivered per year ranged between four and seven. By the late 1990s, the number of ships delivered each year dropped to between zero and four.

Figure 1.3
Number of Combatants Delivered Each Year

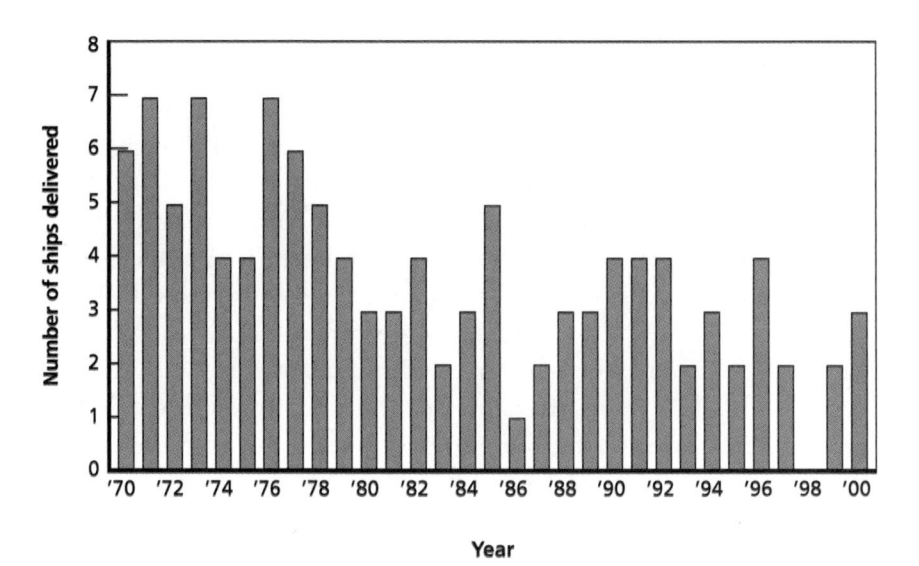

SOURCE: Colton Company, www.coltoncompany.com.
RAND MG294-1.3

The consequence of decreasing ship orders has been a series of closures and consolidations of naval shipbuilders over this period. Figure 1.4 shows a simplistic timeline for the shipbuilders involved in naval and government orders post-privatisation (in the 1970s and 1980s, all the shipyards were nationalised by the UK government and run as part of British Shipbuilders[10]). It should be noted that the figure does not show a complete history of the industry. Hall Russell and Brooke Marine produced MOD ships but went into receivership in 1988 and 1986, respectively. There were also many other commercial shipbuilding firms that were part of British Shipbuilders, e.g., Austin Pickersgill, Doxford and Sunderland, Robb Caledon, Scott Lithgow, and Upper Clyde Shipbuilders (excluding Govan). Few of these firms survive today.

In Figure 1.4, each solid line starts when a shipyard was privatised. If the shipyard subsequently went into receivership, the arrow ends with an 'X' at that point in time. A dotted line represents a facility that was used for or converted to ship repair work only. Mergers are shown as arrows joining.

There are many excellent histories of the UK shipbuilding industry;[11] our intent here is not to provide a detailed or comprehensive history but merely context for the reader. In the mid-1980s, the British government decided to divest itself from the shipbuilding business and began to re-privatise the shipyards. From 1985 to 1990, designated shipyards were sold off. Coincidently, this period also corresponded to the time when naval ship orders began to decline. At the start of privatisation, the naval shipbuilders were, for the most part, profitable.[12] Soon after the privatisation finished, the bottom fell out of the market and these shipyards struggled to survive. There were too many shipyards chasing too few programmes. The intense

---

[10] The exception was Harland and Wolff, which was government owned but not part of British Shipbuilders.

[11] See for example, Johnman and Murphy (2002), Jamieson (2003), Burton (1994), Winklareth (2000), Owen (1999), and Walker (2001).

[12] Johnman and Murphy (2002).

Figure 1.4
History of Naval Shipbuilders Post-Privatisation

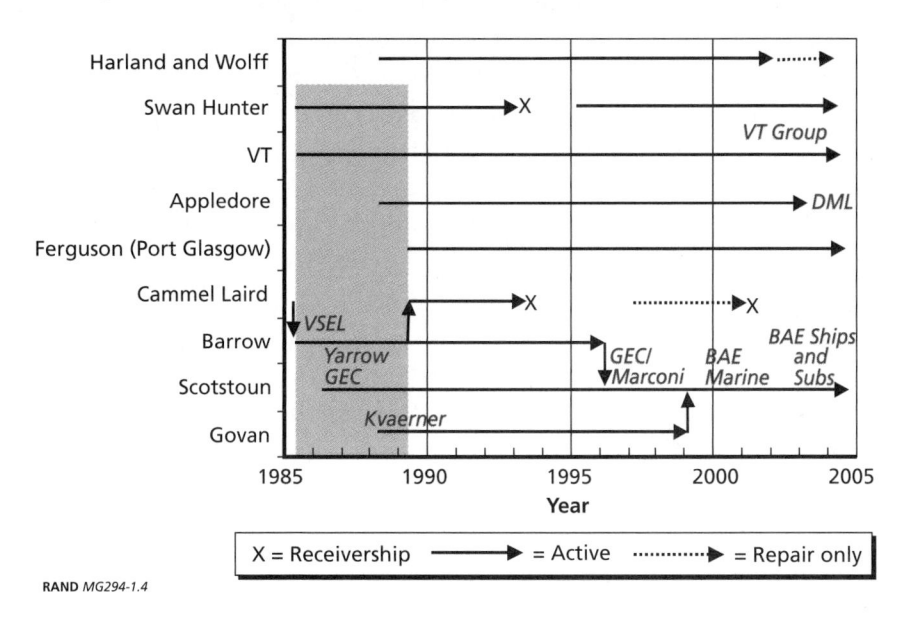

competition that ensued during this period—driven by the MOD policy to compete work—led to very low bids from firms that were simply looking to fill their yards with work. In fact, some have speculated that the bids were, on occasion, below cost.[13] Although this situation may have led to better prices for the MOD, it left the shipyards in a vulnerable state. Certainly, there was little investment, modernisation, or upgrades in the shipyards during this period.

Despite occasional government intervention into the competitive process, Cammel Laird, Appledore, and Swan Hunter shipyards all went into receivership between 1990 and 2004. Some did reopen later. Swan Hunter was reopened and is now the lead shipyard for the new LSD(A) class, and Appledore has been recently purchased by Devonport Management Limited (DML). Harland and Wolff restructured its business in 2002 and is now focusing on ship repair.

---

[13] Johnman and Murphy (2002).

However, it does retain design capabilities and works with other shipbuilders (e.g., NASSCO in the United States). However, its employment levels are greatly reduced from when it was an active shipbuilder.

Other shipyards were consolidated under single ownership. GEC/Marconi consolidated the Barrow and Scotstoun shipyards, which were later sold to BAE Systems. The Govan shipyard was initially acquired by Kvaerner, then sold to GEC/Marconi, and later sold to BAE Systems. By 2004, three major naval shipbuilders were left: BAE Systems, Swan Hunter, and VT Shipbuilding. Ferguson Shipbuilders is also active but focusing more on the coastal patrol, fishery protection, ferries, and other commercial markets. Ferguson could potentially become involved as a producer for future MOD programmes (as either a builder of smaller ships such as the OPV(H) or a producer of blocks or modules on the larger ship programmes).

Issues for Policymakers

The significant future building programme combined with a diminished industrial base raises a number of questions for defence policymakers:

- Is the MOD shipbuilding plan feasible given the constraints of the industrial base?
- What is the programme's effect on the shipbuilders and ship repairers?
- Is the supplier base robust enough to meet the demand?
- Are there alternative timings for programmes that make the plan more robust?
- If procurement quantities change, what will be the effect?

In the autumn of 2003, RAND was asked to address these questions. The analysis described in this report is the documentation of that study. The main goal of this study was to understand the capacity of the UK naval shipbuilding industrial base and its ability to undertake the MOD's shipbuilding programme over the next 15 years.

There are two issues not addressed in this document: the affordability of the shipbuilding plan and whether the plan meets the customers' requirements. By affordability, we mean the ability of the government or the DPA to budget for ship acquisitions. An increase in procurement activity will need, generally, to be supported by an increase in funding. By customers' requirements, we mean the ability to meet the operational needs in terms of both timing and capability. The scope of this study was to look at issues from an industrial perspective and not from a funding, political, or an operational needs perspective. However, any long-term plan will need to balance these issues with the industrial base ones.

Study Structure

Assessing the capacity of the entire naval shipbuilding industry is a very complicated problem. As discussed earlier, there are many firms involved in the industry, ranging from very large conglomerates to small businesses with fewer than 50 employees. These firms serve as prime contractors, shipbuilders, ship repairers, equipment suppliers, service providers, and designers and architects. Therefore, a complete analysis must touch on the spectrum of firms involved. Furthermore, capacity to manufacture cannot simply be stated in simple terms, such as a specific number of ships per year. Production capacity very much depends on both the types of ships produced and the build strategy employed. For example, a much higher rate of frigate production is possible compared with the rate for aircraft carriers at a given facility.

To analyse the issues the MOD faces, we decomposed the capacity evaluation into a supply and demand assessment in three distinct areas: labour, facilities, and suppliers. Figure 1.5 (the study's hierarchy) shows that a specific shipbuilding plan results in a series of demands on the industry. Although each area is, to a certain extent,

**Figure 1.5
Study Hierarchy**

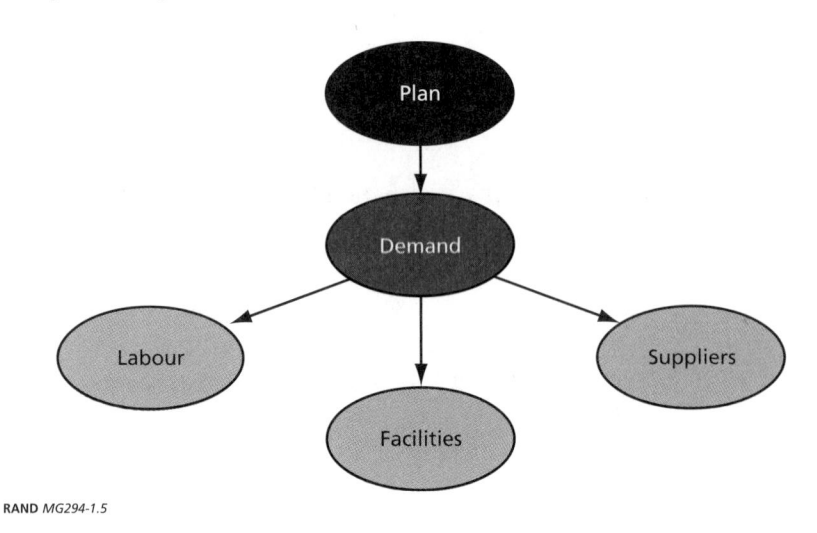

Diagram: Plan → Demand → Labour, Facilities, Suppliers

RAND MG294-1.5

independent, each area is connected to the plan. If one area has insufficient capacity, the plan becomes problematic. Often the result of insufficient capacity are schedule delays and cost increases.

Plan

The plan is the 'who, what, and when' of the shipbuilding programme. As described earlier, many of the programmes are in the early stages of definition and planning (pre–Main Gate). Therefore, it is not possible to have a conclusive allocation of work to the shipyards, schedule of activities, and set of requirements.

So rather than focusing on a single future, we will present several alternative plans. Each plan will be defined in greater detail in the subsequent chapters. Having several alternatives will make the study more robust as MOD plans evolve. Having alternative futures also will highlight sensitivities of various assumptions. The study will use as its baseline the 'current plan', which assumes that everything is built as envisioned by the IPTs and Equipment Capability Customers (ECCs). Another plan that we will explore is one in which funding or requirements decrease such that fewer vessels are purchased (pessimistic funding). We also examine the plan in which requirements or funding increase (optimistic funding). We will examine the scenario where a new, large submarine (similar sized to the *Vanguard* class) is designed and built. In addition, we will present a plan in which timings are stretched out (level-loading).[14]

Labour
The demand for labour is viewed, at least by the shipyards, as the most critical issue they face in meeting the MOD's plans over the next several years. From a labour perspective, policymakers need to understand both what labour is required and what labour is available. However, labour is not a static commodity: Workers retire; apprentices enter the workforce; workers change employers; etc. Thus, it is crucial to understand the dynamic of how demand and supply of labour evolve over the next several years. To address the complex issues of labour, the study team split the analysis into two parts. Chapter Two will examine the demand for labour under the different plans described above. Chapter Three will examine the supply of labour and whether it matches the demand.

Facilities
As described at the beginning of this section, shipbuilding requires substantial facilities, which are expensive to create and are generally spread over a large footprint. Hence, expanding or upgrading facilities cannot be done quickly. Thus, in addition to labour, decision-makers will need to understand the availability of facilities to meet the demand for shops, piers, and docks. Chapter Four will look at the analysis on these last two types: docks and piers.

Suppliers
More than half the value (unit cost) of a naval vessel is provided by firms other than the shipbuilder.[15] So the ability of suppliers to meet

---

[14] One further scenario is presented in Appendix A on the implications of programme delay.
[15] *The Clyde Shipyards Task Force Report* (2002, p. 41).

the demand based on the MOD's plans is an important consideration in addressing the UK industry's capacity. Chapter Five explores supplier issues such as stability, diversity of business, and ability to meet increased demand.

Survey of the UK Shipbuilding Industry

Doing an encompassing study such as this entails interviewing and surveying a great number of firms and groups. The MOD and RAND agreed to a list of firms to contact. Some of these firms chose not to take part in the study. Those that did participate were

- AMEC
- Atkins Global
- BAE Systems (Naval Ships and Submarines)
- Babcock Engineering Services (BES)
- British Maritime Technologies (BMT)
- Devonport Management Limited
- Ferguson Shipbuilders
- Fleet Support Limited (FSL)
- Kellogg-Brown and Root (KBR) Caledonia
- Swan Hunter
- Thales UK
- VT Shipbuilding.

The RAND team sent surveys to each of these firms requesting information on their employment, future workload demands, facilities available, and their key suppliers. The team then did follow-up interviews with the firms to clarify the data and to discuss any issues of importance that were not covered by the surveys.

The study team next contacted the key suppliers that the firms had identified. (The shipyards identified more than 200 suppliers, which will not be listed here because of space limitations.) The team sent a separate survey to these suppliers asking them about their employment, relative competitiveness of their market, their dependence on MOD and maritime work, and the challenges they will face in the future. There were occasional follow-on conversations with the suppliers that did return the survey—mostly for clarification.

In addition, the RAND team interacted with a number of industry associations:

- Furness Enterprise Limited
- Highlands and Islands Enterprise
- Northwest Development Agency
- Scottish Enterprises
- Sector Skills Council for Science, Engineering, and Manufacturing Technology (SEMTA)
- Society for Maritime Industries (SMI)
- Shipbuilders & Shiprepairers Association (SSA).

Finally, the team corresponded with several government agencies:

- Department of Trade and Industry (DTI)
- DPA
  –All Naval IPTs
  –Future Business Group (FBG)
  –Pricing and Forecast Group (PFG)
  –Sea Technologies/Quality Group
  –Supplier Relations Group (SRG)
- MOD Directors of Equipment Capability (DECs).

These government interactions mainly involved interviews and the request for specific data. For example, the PFG provided notional production hours for pre–Main Gate programmes. Each IPT provided notional timings for its specific programme.

Study Outline

We divide the report into seven chapters. Chapter Two discusses labour demand issues. Chapter Three examines labour supply and the ability to meet the demand. Chapter Four analyses the facility implications of the various plans. Chapter Five describes and discusses the suppliers. Chapter Six explores the resources available at medium-sized shipbuilders and commercial firms. Chapter Seven summarises the study and lists issues (beyond capacity) for the MOD to consider as it defines the shipbuilding strategy for the United Kingdom. Lastly, Appendix A shows how schedule slippage affects MOD labour demands; Appendix B provides a short reference of ship dimensions; and Appendix C breaks out skills by management/technical and manufacturing categories.

CHAPTER TWO
Labour Demand

In this chapter, we take a closer look at the labour demand that the MOD's current and future shipbuilding programme will place on the shipbuilding industrial base. Although issues concerning the current shipbuilding skill base and facilities will be examined in later chapters, it is critical for the MOD to clearly understand both how much and what types of demand it may place on the industrial base through its planned shipbuilding programme. A proper understanding of this impact is key in determining whether existing resources will be able to meet the projected workload under the appropriate timelines.

Because many of the MOD's future programmes have not been specifically defined, it is necessary to develop projections that will allow us to estimate and assess the impact of these programmes on the industrial base. Over the next several pages, we will lay out the methodology we developed to assess future MOD-generated labour demand, explain the basic data we used to assess this demand, and present the assumptions we used in our demand estimates. After presenting the results of our examination of the MOD's current acquisition plan and its implications for the industrial base, we will also briefly examine a number of alternative scenarios. Finally, we will consider ways in which the MOD could attempt to level future demand to make it easier for the industrial base to cope.

Methodology

The goal of our labour projection model is to estimate future labour demands for the MOD's shipbuilding programme. To do that, we followed a straightforward process. First, we developed a labour projection model that would allow us to estimate future demand. Second, we made a set of basic assumptions concerning the future MOD shipbuilding programme, including the timing and number of ships in a particular planned class (e.g., CVF, MARS, FSC). Third, we collected data about these programmes and populated our labour projection model with that information. Finally, we ran the model to produce a number of estimates concerning the future labour demand placed by MOD programmes. Because the future naval procurement plan is flexible and open to change, we also ran a number of other scenarios to look at the robustness of our projections. The first step in our methodology was to develop a model that can accurately forecast future labour demands.

The model used to project future MOD labour demand required a number of inputs. We show the model's general formulation in Figure 2.1.

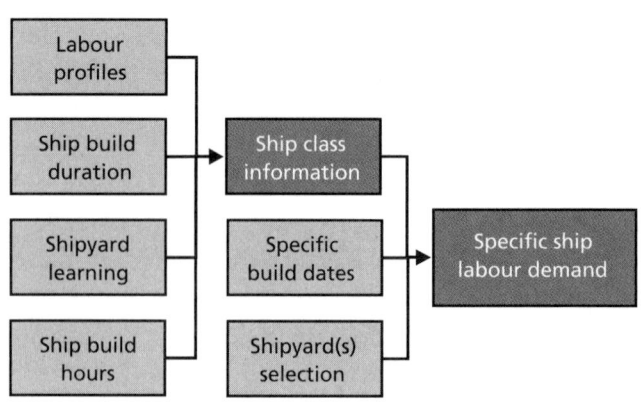

Figure 2.1
RAND's Basic Labour Forecasting Model

The forecasting model requires data at a number of different levels. First, there is specific information associated with each ship class (e.g., Type 45 or CVF). Such class-specific information includes:

- *Labour profiles.* These profiles represent the distribution of labour required to build and design a ship over time. Although there may be a generic labour profile required for the overall production of a ship, it is also possible to estimate labour profiles for the individual skill trade areas associated with building a ship. In our model, we developed labour profiles unique to skill trade categories (which we will define later in this chapter).
- *Ship build hours.* These are the total number of direct worker hours required to complete a ship, which can be further broken down by individual skill trade areas (i.e., management, technical, structural, support, and outfitting).
- *Ship build duration.* This duration represents the total amount of time needed to actually build a ship. At this point in the modelling process, we did not identify specific start or stop dates for ship design or production, but instead made a general estimate of the overall block of time needed to build the ship (e.g., 10 quarters).
- *Learning curves.* These curves represent the labour hours required over time as more ships of the same class are produced. This improvement can come through increased labour proficiency, process innovation, or some combination of the two. Typical unit learning curve slopes[1] for shipbuilding range from 0.87 to 0.93, and we used different learning curves depending on the ship class under consideration.

Past RAND studies have used this direct labour estimation model to look at overall labour demand. However, for this study, we

---

[1] By *unit learning curve slope*, we mean the rate of improvement each time the production doubles. A 0.95 slope, for example, means that the hours decrease by 5 percent each time the production unit doubles. It is a nonlinear improvement, getting smaller as the production quantity increases. See Arena, Schank, and Abbott (2004) for more details.

broke down overall labour into more definitive categories so that it would be possible to analyse labour demand at each of these levels. The five categories of labour skill levels we looked at were:

- Management: general management, administration, marketing, and purchasing
- Technical: design, drafting, engineering, planning, and project control
- Structural: steelworking, structural welding, shipwright, etc.
- Outfitting: electrical; joinery; heating, ventilation, and air conditioning (HVAC); insulation; painting; etc.
- Support: rigging, scaffolding, storekeeping, cleaning, etc.

Appendix C details a more exhaustive list of specific skills that we included in the general skill trades.

It is also important to note that the labour estimation model projects only direct labour and does not account for overheads that will be unique to individual shipyards, prime contractors, and design firms. For our purposes, we define direct labour as labour specifically charged to a particular project by industry.

Thus for an individual ship class, it would then be possible to create representative, time-independent labour profiles for the number of direct workers required to design and produce a ship over time. We present a notional example of such profiles in Figure 2.2.

However, to convert these notional labour projections into an actual estimate, two other important pieces of ship-specific information are required: specific start and end dates for production need to be defined, and a specific shipyard(s) must be selected to actually design and produce the vessel. Only when these two steps are completed will the nominal labour projection shown in Figure 2.2 be useful in estimating future labour demand for a specific naval vessel.

In our model, we made estimates for both these pieces of information. (We cover the specific information later in this chapter.)

Figure 2.2
Example of Direct Labour Distribution Curves for an Individual Ship Class

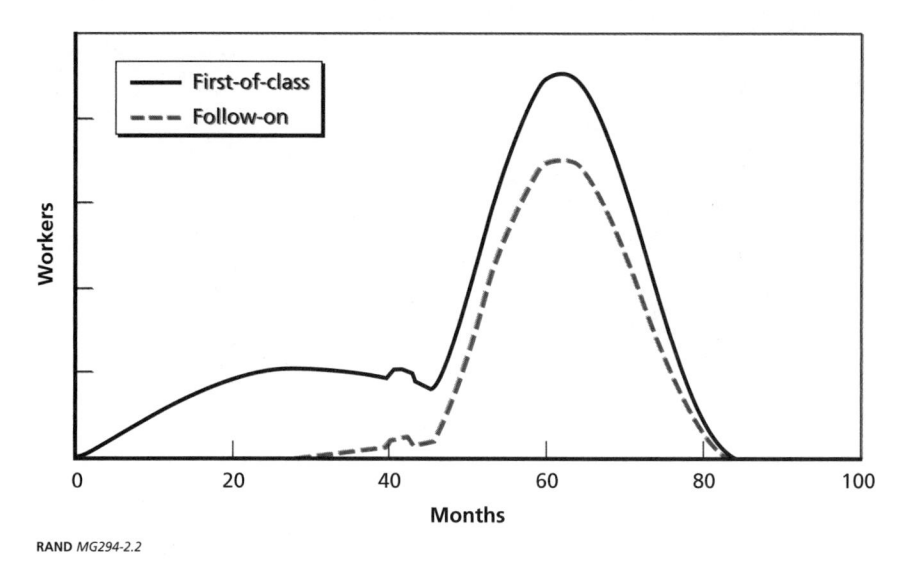

As estimates are made for each future ship for the Royal Navy, it becomes possible to aggregate these individual estimates to come up with an overall labour estimate for the entire MOD shipbuilding programme. As mentioned before, these labour estimate aggregations can then be analysed at either a programme, skill trade, or shipyard level. We provide a visual representation of this aggregation in Figure 2.3.

Basic Assumptions

To apply this basic model of direct labour estimation to the MOD's current and future shipbuilding programme, we needed to make assumptions about what that programme may look like. In addition to making assumptions about the general characteristics of each

Figure 2.3
Individual Ship Aggregation to Represent Entire Shipbuilding Programme

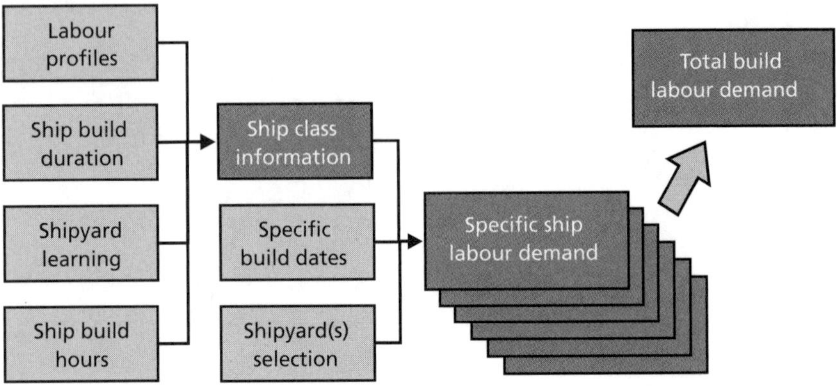

RAND MG294-2.3

future Royal Navy ship class (build duration, build hours, etc.), we also needed to estimate the number of ships in each class and the timings of the design and build for each specific ship.

Recognising that it would be difficult to accurately project all this information, we defined one set of assumptions as our current MOD plan, but then ran a number of alternative scenarios to understand the labour demand impacts of programme variations. In the current MOD plan, we included all MOD ships currently being built as well as estimates for future ships likely to be built. Additionally, we included some ship repair/refit projections (as envisioned by the repair yards) so that the MOD could consider the impact of its repair and refit programme on its new-build programme.

The current MOD programmes considered in our analysis were

- Type 45: the first six ships
- Astute: the first three submarines
- LSD(A)
- LPD(R).

These ships are currently on contract, are or have been designed, or are in production. We provide a summary of their projected status in Table 2.1.

Of the future MOD programmes, we included

- Astute (the fourth submarine and beyond)
- CVF
- Future Minehunter
- OPV(H)
- FSC
- JCTS
- MARS
- Type 45 (the seventh ship and beyond).

We had to make assumptions regarding both the number of ships in each class and the time period for the design and production of each class (Table 2.2).

Table 2.1
Current MOD Shipbuilding Programmes

Programme	Class Size	Design/Build Duration
LPD(R)	2	Through 2004
LSD(A)	4	Through 2006
Type 45	6	Through 2010
Astute	3	Through 2011

Table 2.2
Projected MOD Shipbuilding Programmes

Programme	Class Size (on contract, additional)	Design/Build Duration (from 2004)
OPV(H)	0,2	2004–2006
JCTS	0,1	2006–2010
CVF	0,2	2004–2015
Astute	3,4 (7 total)	2005–2018
Type 45	6,3 (9 total)	2004–2012
MARS	0,10	2007–2020
FSC/Future Minehunter	0,14/4	2007–2023

We independently made a number of additional assumptions about how each ship would be built (modular build, split-location production, etc.) and where it would be built after discussions with MOD officials (IPTs, other DPA staff, and ECC representatives), design firms, shipbuilders, and other stakeholders. They represent a reasonable, but not definitive, projection as to what the future MOD shipbuilding programme could look like. Because the shipbuilding programme may change over time, we look at a number of alternative shipbuilding projections later in this chapter to show the impact to the industrial base of changing assumptions.

Although the information presented in Table 2.2 outlines our baseline assumptions regarding the MOD's future shipbuilding programme, it is appropriate to discuss each programme briefly to outline further ship class-specific assumptions.

- *OPV(H)*. We assumed that these ships would be of the same size as the *River*-class OPVs and would be built as whole ships in one location using traditional shipbuilding methods.
- *JCTS*. We assumed that the JCTS would be of slightly larger size than the LSD(A) ships currently in production and would be built using similar methods, with the entire ship being built in one location.
- *CVF*. We based our assumptions for the CVF on a traditional modular production plan in which large blocks of the carrier would be constructed at several shipyards around the United Kingdom, eventually being transported to a final site for assembly and integration.
- *Astute*. We assumed that these submarines would be built using the same production methods as the first three in the class, but that there would be incremental upgrades applied for the fourth submarine and beyond that would slightly increase required labour as well as lead to additional nonrecurring design work for the fourth Astute.
- *MARS*. We assumed that the MARS programme would encompass 10 ships, which would be further broken down into smaller class groups. The exact number and size of each group has not

yet been determined, but our assumptions took into account the most-current thinking of the IPT that the displacement of each class would vary but would be largely in line with the size of current RFA supply ships. We also assumed that an individual MARS hull would be built in one location, but that multiple locations would produce hulls.

- *Type 45.* We assumed that future Type 45 destroyers would be built in the same modular way as the current batch with no significant capability upgrade in the last batch of ships.
- *FSC.* Although there are a number of potential designs being considered for the FSC, we assumed that the ship would be a mono-hull, steel structure, which is similar in size to the current Type 23 fleet. We further assumed that the individual ships would be constructed at a single location.
- *Future Minehunter.* We made a deliberate assumption that the Future Minehunter would not be like the current fleet of minehunters but would instead operate as a central platform with remote unmanned underwater vehicles responsible for mine detection. This assumption reduced the number of platforms required. For simplicity, we assumed that this platform would be of the same basic form as the FSC and would be constructed in the same way.

Additional Assumptions

In addition to the assumptions above, we considered projected MOD repair/refit work in our labour projections, which were based on information received from the three warship repair/refit yards and the MOD. We also included projections for both non-MOD work (shipbuilding-related or not) and military export orders, where we judged the probability of success to be reasonable. Our refit work estimates included labour demand estimates from two of the three major MOD refit yards.[2]

---

[2] The third major refit yard provided us with general employment data and met with us to discuss its plans, but it declined to provide future labour projections because of economic sensitivities. However, we are confident that our refit labour projections captured the

There were a number of potential MOD programmes that we did not include in our analysis. We assumed these ships would not be new-builds and either may not be replaced by the MOD as they leave service or would be taken up from trade and refitted (and thus would not tax the shipbuilding industrial base). The most high-profile ships that we assumed would not be new-builds and did not include in our analysis were the following:

- *RFA* Argus *(training flight deck role).* The medical role of the *Argus* will be taken up by the JCTS. We assumed that the air training role of the ship would either be replaced by simulation, taken up by current ships in the fleet, or some combination of the two.
- *RFA* Diligence. We assumed that a replacement for this vessel would be taken up from trade and refitted.
- *Additional helicopter landing platform (LPH).* Should an additional LPH be required, we assumed the one of the CVS vessels would get a refit/life extension to fill this role.
- *Landing craft.* Although a number of these small craft may be replaced, they place a negligible load on the industrial base, and thus we did not model their demand.

Once we had defined our model and assumptions, we collected a wide variety of data to populate it and produce the results. The data we needed roughly corresponded with the model inputs defined earlier in this chapter. To collect these data, we went to a variety of sources, including traditional military shipbuilders and ship repairers, commercial shipbuilding firms, naval prime contractors, design houses, offshore marine firms, and trade associations. Where appropriate, we followed up with a visit to the firm. We also visited with individual shipbuilding IPTs, their customers in the ECC, and the DPA's SRG and PFG. At times, the data we received were inconsis-

---

majority of future refit demand and that the absence of data from the third major refit yard did not affect any future MOD refit demand trends (which show a general decrease as the fleet size shrinks).

tent, and we were required to seek further clarifications and make informed judgements to reconcile them. Additionally, because some of the programmes surmised are still in the concept or preconcept phase of the acquisition cycle, our estimates for these programmes will be open for further refinement as more information about them is made more concrete.

Current MOD Plan: Overall Labour Demand

Figure 2.4 shows the expected direct labour demand that the MOD's shipbuilding programme will place on the United Kingdom's industrial base over the next 20 years, based on our assumptions and the data we collected.

**Figure 2.4
Future MOD Labour Demand, by Programme**

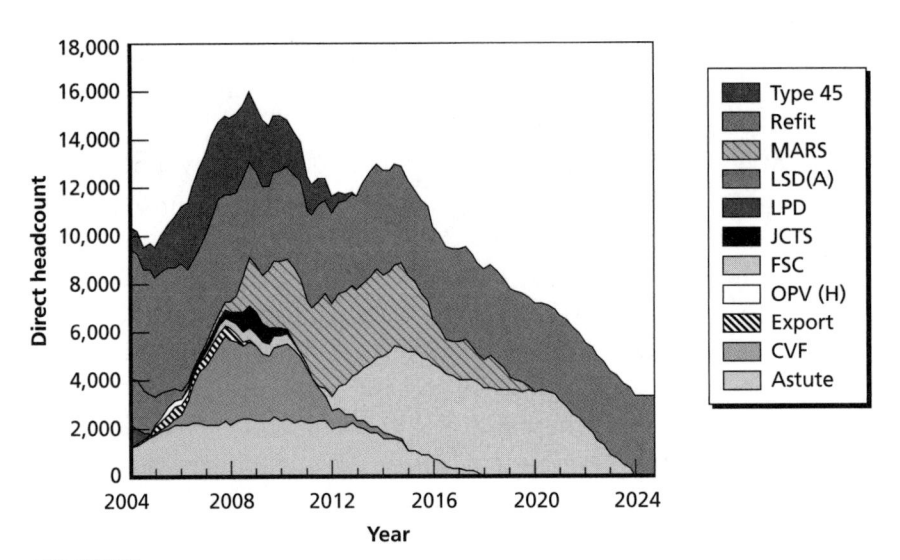

As the figure shows, the MOD's labour demand at the beginning of 2004 is slightly more than 10,000 direct workers. This demand comes from four major programmes (Type 45, LSD[A], LPD[R], and Astute) as well as refit work. This future demand will decrease slightly as the LSD(A) and LPD(R) programmes end, temporarily dropping to slightly more than 9,600 workers in early 2005. However, as further MOD programmes begin to move into their design and production phases, demand will increase significantly. This increase is largely a result of the CVF programme, but it also takes into account JCTS design and production, MARS design, and FSC design. As the MARS programme begins to move into full production, demand will peak at slightly more than 16,000 direct workers at the end of 2008. At peak, four major programmes—Type 45, MARS, CVF, and Astute—will be in production, and the JCTS and FSC programmes will also be placing additional labour demands on the industrial base. After this peak, labour demands remain above 14,000 workers through 2010, whereupon production on the first CVF will begin to decrease, which will lead to a steep drop in labour demand. Still, overall labour demand will continue to average close to or above 12,000 direct workers through 2015, when the MARS programme begins to wind down. At this point, the only major MOD programmes still in production (other than refit work) will be MARS, FSC, and Astute. Once the MOD's labour demand drops below 12,000 direct workers, it will steadily decline into the future. It is possible that the MOD may embark on additional shipbuilding programmes during the 2015–2020 time frame, but we considered this possibility unlikely (except for a potential Future Submarine that we take into account in an alternative scenario later in this section).

Concerning the global outlook of these demand requirements, the MOD's future programmes will increase their demand on the industrial base by more than 50 percent above current MOD demand. This increased demand takes place over a five-year period, with a longer period following when the demand will decline more steadily.

Current MOD Plan: Demand for Specific Labour Skills

It is also possible to look at future demand placed on the industrial base at the general skill levels defined earlier in this chapter. An analysis at this level will allow the MOD to better understand the specific impacts as different skills demands stress the industrial base at different times.[3]

Management Labour Skills

First, we consider the demands placed on management skills. Figure 2.5 shows the projected management skill level demands placed on the industrial base by the current MOD ship procurement plan.

Figure 2.5
Future MOD Labour Demand for Management Skills, 2004–2025

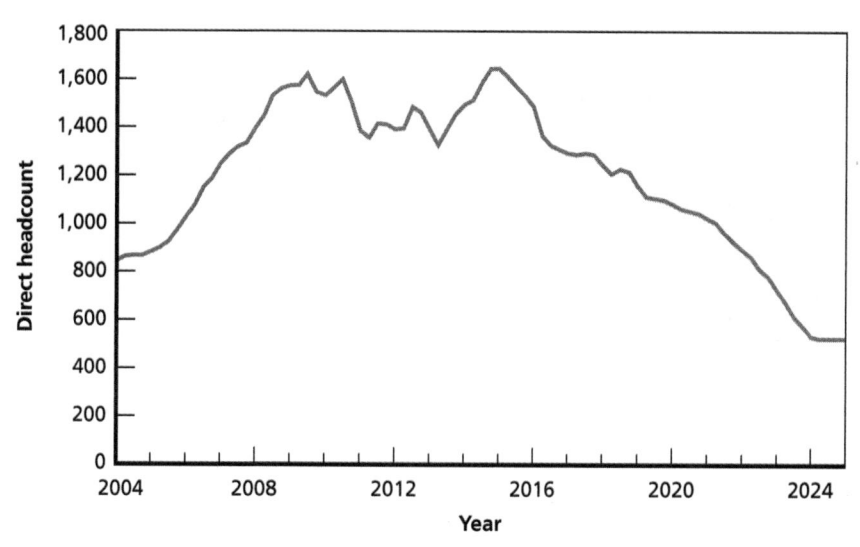

RAND MG294-2.5

---

[3] For example, for an individual ship, structural skills are generally required earlier in the production cycle than outfitting skills are. Additionally, technical skill requirements will be used even earlier in the cycle to complete the design prior to the start of production.

This figure shows that overall management demand is expected to rise steadily over time from its current state of approximately 850 workers, almost doubling between 2004 and 2009. The demand then stays fairly constant for the next six to seven years before slowly returning to its current levels by 2022.

Technical Labour Skills

Next we consider the demand placed by the MOD on technical workers in the industrial base (see Figure 2.6).

This projected technical labour demand shows that the MOD demand for technical labour is expected to decrease by approximately 600 workers in the next year from its current level of more than 2,700 workers. However, because of the future design requirements for the CVF, MARS, and FSC programmes, this type of demand increases sharply until reaching a peak level of slightly less than 3,800

Figure 2.6
Future MOD Labour Demands for Technical Skills, 2004–2025

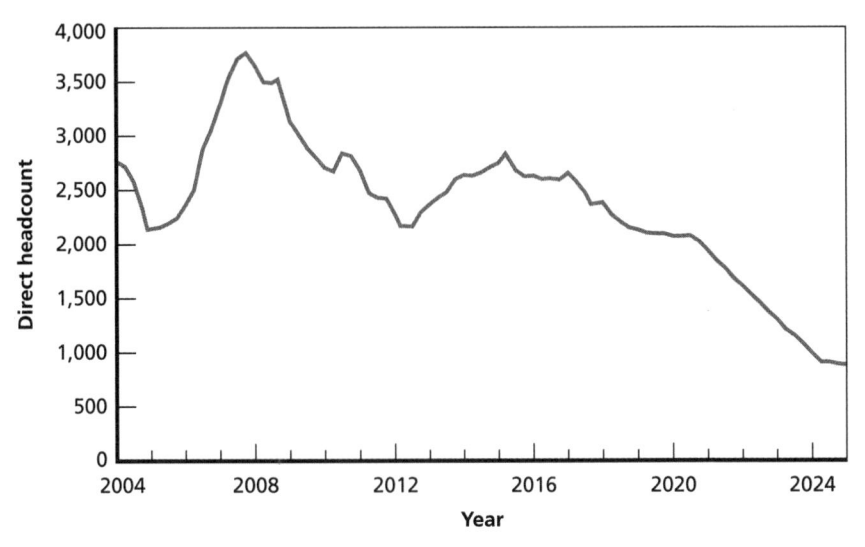

RAND MG294-2.6

workers in mid-2007. The demand then decreases at a somewhat uniform rate until 2012, whereupon it remains fairly constant until 2020. However, it is the initial immediate decrease in technical demand followed by a sharp upturn that should concern the MOD. Should the shipbuilding industry not retain the technical workers during the downturn, the industry will face the need to recruit more than a thousand new workers in a two-year period.

Structural Labour Skills

Figure 2.7 shows the demand placed on structural skills by the MOD, revealing that the MOD's future naval programmes will place a steadily increasing demand on structural skills over the next six years, culminating in a structural worker demand in excess of 3,300 workers (an increase of more than 60 percent) before sharply tailing

Figure 2.7
Future MOD Labour Demand for Structural Skills, 2004–2025

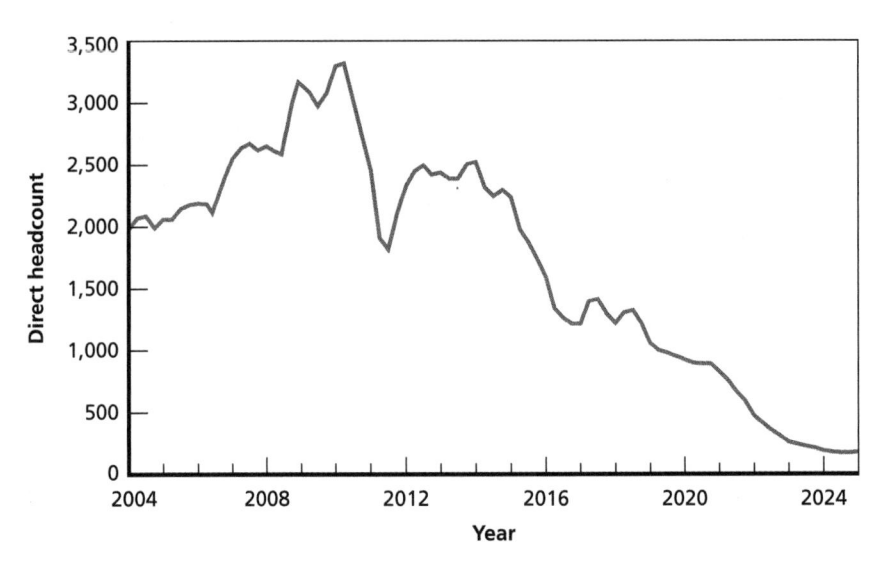

RAND MG294-2.7

off after 2010. Thereafter, demand for structural workers will rise slightly for five to six years before steadily decreasing.

Outfitting Labour Skills

Next, we look at the demand placed by the MOD for outfitting skills (Figure 2.8).

After staying fairly constant until 2005, the outfitting demand then begins to increase at a sharp rate, almost doubling over a five-year period from just over 3,200 workers to slightly more than 5,800 workers. As with the overall demand, this outfitting demand thereafter decreases steadily over the next 15 years, with only a slight increase during the 2012–2015 time frame. The outfitting skill trade is the largest of the five skill trade categories we defined, and the shape of its curve mirrors that of the overall labour demand curve.

Figure 2.8
Future MOD Labour Demand for Outfitting Skills, 2004–2025

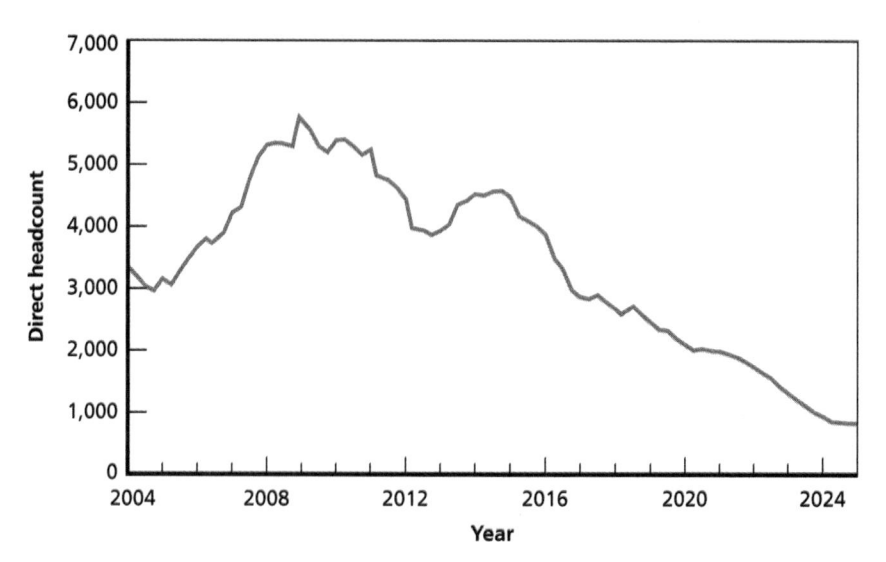

RAND MG294-2.8

Support Labour Skills

The final skill trade that we consider in our analysis is support. Figure 2.9 shows the demand on this skill over the next 20 years.

Of all the skill trades considered, support has the least variation over time. Still, the demand for support skill trades is expected to increase by approximately 33 percent over the next five years and stays at or above current levels until 2015, after which the demand slowly decreases.

Macro Versus Micro View of Demand

Thus far, we have looked at the MOD's projected labour demand both at the overall, general level, taking specific programmes into account, and at the skilled trade level (i.e., the macro view). At each

Figure 2.9
Future MOD Labour Demand for Support Skills, 2004–2025

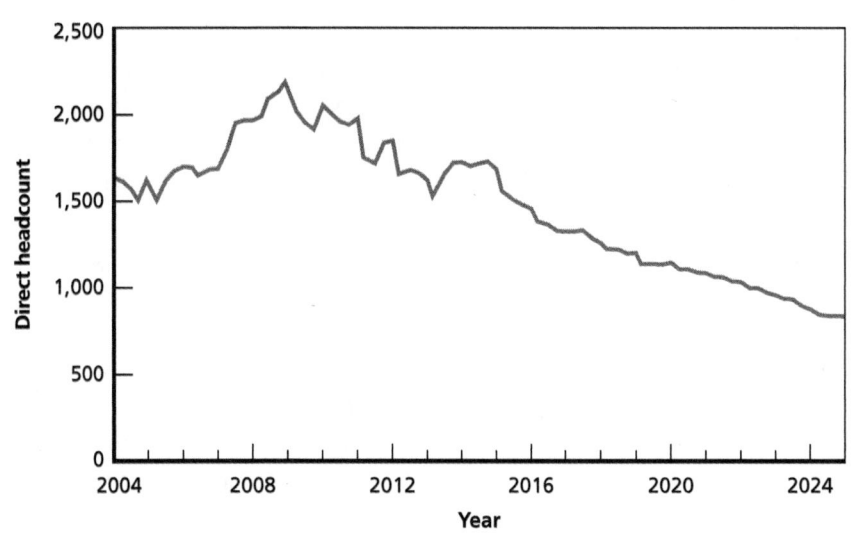

level, we have identified trends in future demand, and we will later consider the implications of these trends. We also conducted this analysis on a shipyard level, assigning shipyards to build specific programmes and analysing the impacts of the overall programme on specific yards (i.e., the micro view). However, given both the commercial sensitivities surrounding this issue and the uncertainly over which firms the MOD will select to design and build specific ships, we have omitted that part of the analysis from this report. However, the importance of this shipyard-specific analysis should not be underestimated. The aggregate (macro view) analysis of MOD demands placed on the industrial base will be able to help the MOD understand the effect of its programme on the overall industry. It is also important for the MOD to know the impact of its programmes at an individual shipyard level, because excess or insufficient demand could cause productivity, economic, facilities, or other strains on individual yards.

Furthermore, the macro view implicitly assumes that work can be shifted freely between facilities and, in essence, supposes a single labour resource pool to do all the work. This situation, of course, is not true. There are real constraints (social, practical, and procedural) to the extent that work and workers can be shared. For example, work during the later stages of outfitting must be done on the ship not in a shop or assembly area. Thus, this particular work is constrained to the location where the ship is docked or berthed. The work cannot be subcontracted to another location. Similarly, workers cannot move readily between shipyards—say, for example, to be in the northeast one day, southern England another, and Scotland the next. We are not implying that the shipyards do not or cannot share work. In fact, they have done so and do now. VT is a subcontractor to BAE Systems on the production of the Type 45. Appledore produced the *Echo*-class survey vessels as subcontractor to VT. Swan Hunter subcontracted sections of the LSD(A) to other fabricators in the northeast. Sharing of work between the firms will help to level the labour demands at a micro level, but it will not reduce the overall demand (macro view).

The real implication of not fully being able to share labour resources is that the actual employment levels might need to be higher than the macro level demand shows.[4] This situation arises, as any particular shipyard might need to employ a larger workforce to meet future peak demands. Shipyards cannot, in general, expand and contract their workforce without limit to meet demand. It takes time to find and recruit workers. Often, newer workers must be trained by more experienced ones. Expanding the workforce too rapidly may result in lower productivity. Therefore, the MOD will need to consider the demand it places on the firms on a firm-by-firm basis to fully understand the implications of its build plan. In the next chapter, we will examine the implication of constraints to expanding the workforce.

Alternate Future Scenarios

Thus far, our analysis has focused on the demand for labour presented by the MOD's current ship acquisition plan. However, because this plan may change in both scope and timings of programmes, we looked at three alternate scenarios involving changes in our basic assumptions. These scenarios allowed us to examine the degree of sensitivity associated with the current programme and identify areas that the MOD may need to closely monitor should its future plans change.

The first scenario we explored looked at what would happen to future MOD demand should the number of ships in the future programme be reduced, either by decreased requirements or insufficient budgets. The second scenario examined the impact of adding a Future Submarine to the assumed future programme. The third scenario looked at the impacts of increasing the future requirements for the Royal Navy, which would manifest itself in increased ship orders.

---

[4] For more information on why the actual employment might be higher, see Arena, Schank, and Abbott (2004).

Scenario 1: Decreased MOD Requirements or Budgets

In this section, we examine the implication on the future MOD shipbuilding demand if requirements for future naval requirements were cut. In Table 2.3, we represent these decreases from the current MOD plan in the shaded sections.

As Table 2.3 shows, we assumed that in a future with decreased requirements there would be no additional Type 45s built, the total number of Astute submarines and MARS ships would each decrease by one ship, and there would be two fewer Future Minehunters built. To compensate for fewer Type 45 destroyers, the start date for the FSC build was moved forward.

When we put this amended data into our future labour demand model, we found that the overall MOD demand on the industrial base would likely decrease substantially. We show this decrease in Figure 2.10, which depicts the demand level from the MOD's current plan as a dotted black line.

Overall, the peak demand generated by this scenario would be lower than the current MOD plan, which is to be expected because this scenario had fewer ships being built. As with the current MOD

Table 2.3
Scenario 1: Decreased MOD Requirements or Budgets Programme Assumptions

Programme	Class Size (on contract, additional)	Delta Number of Ships from 'Current Plan'	Design/Build Duration (from 2004)
OPV(H)	0,2	0	2004–2006
JCTS	0,1	0	2006–2010
CVF	0,2	0	2004–2015
Astute	3,3 (6 total)	−1	2004–2016
Type 45	6,0 (6 total)	−3	2004–2010
MARS	0,9	−1	2007–2020
FSC/Future Minehunter	0,14/2	−2	2007–2022

Figure 2.10
Scenario 1: Decreased MOD Requirements or Budgets—
Labour Projections by Programme, 2004–2025

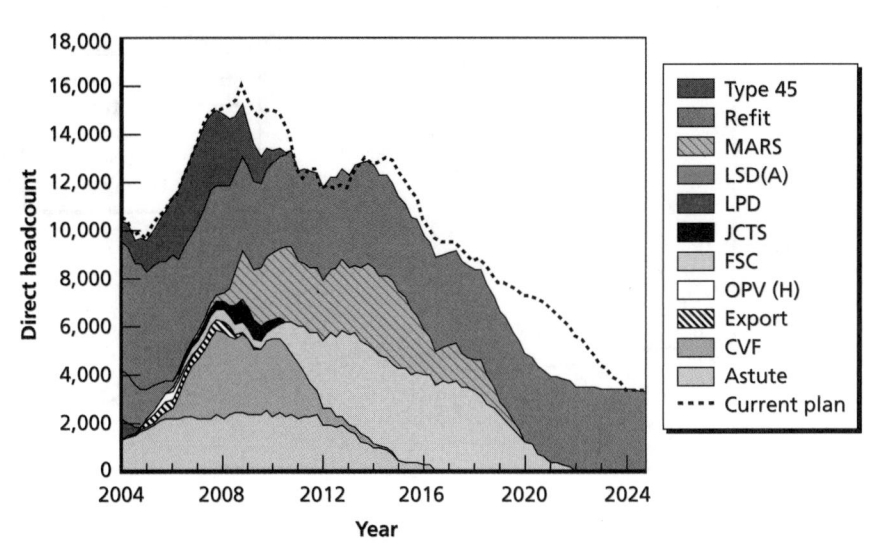

RAND MG294-2.10

plan, there is a slight decrease in demand in 2005, followed by an upsurge in overall demand. Although this demand peaks at the same time as in the current MOD plan, that peak involves about 800 fewer workers than the MOD plans. Other than a slight increase over the base case, due to the FSC being moved forward, the first scenario generally follows the general demand pattern of the current MOD plan but with less labour demands. The first scenario diverges from the MOD plan in the later years as its future surface combatant production drops off.

We also looked at the first scenario's labour demands by skill level. Figure 2.11 shows the MOD's direct labour demand using assumptions from the first scenario.

**Figure 2.11
Scenario 1: Decreased MOD Requirements or Budgets—
Labour Projections by Skill Trade, 2004–2025**

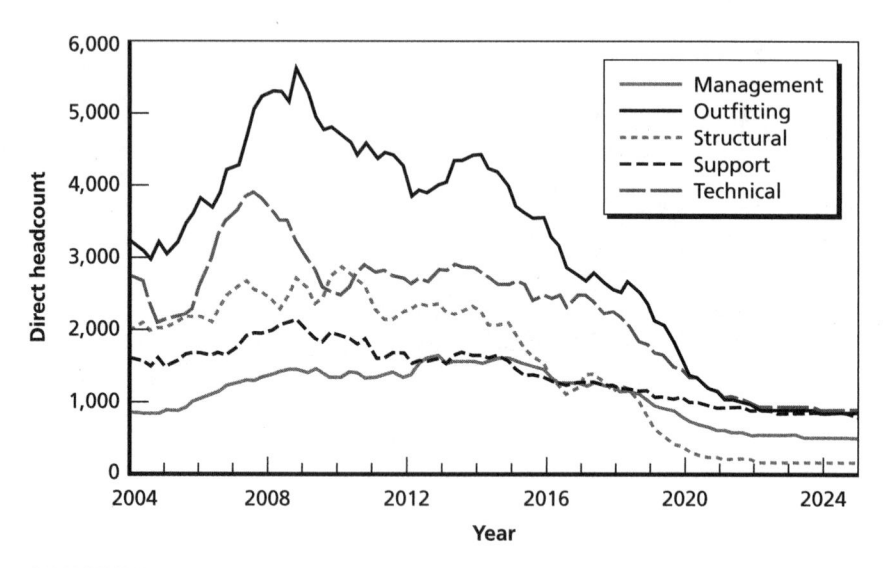

The general skill level labour demand trends are similar to those in the current MOD plan. However, there are a couple of important differences. First, the maximum demand placed on outfitting skills by the first scenario is less than that of the MOD plan (although both cases do require a rapid increase in outfitting labour from 2004 to 2009). The other important difference is in the technical demand. Although, again, the total number of technically skilled workers decreases in this scenario, the 'hump' of technical demand is more pronounced in this case.

Scenario 2: Addition of Future Submarine to the MOD's Requirements

In this scenario, we considered the impact of adding a Future Submarine to the MOD's future requirements. Because of the sensitivities regarding the role such a submarine may fulfil, we did not define the function of the submarine, but only assumed that it would be

Vanguard-sized, built to a modified Astute design, and enter into service around 2023.[5] In Table 2.4, we highlight this assumption in the shaded sections.

Inputting this additional information into our model, we were able to project the labour demand that this Future Submarine may place on the industrial base. Figure 2.12 shows the overall MOD demand with the demand for a Future Submarine in at the top in black with white stripes. As before, the demand placed by the current MOD plan is shown as a dotted black line.[6]

As we assumed that the design and production of a Future Submarine would start right at the end of this decade, this scenario has the same peak demand as the current MOD plan. However, the addition of a Future Submarine significantly increases the demand on the industrial base from 2009 onwards. This Future Submarine also cushions any decrease in demand into the 2020s.

Table 2.4
Scenario 2: Addition of Future Submarine Programme Assumptions

Programme	Class Size (on contract, additional)	Delta Number of Ships from 'Current Plan'	Design/Build Duration (from 2004)
OPV(H)	0,2	0	2004–2006
JCTS	0,1	0	2006–2010
CVF	0,2	0	2004–2015
Astute	3,4 (7 total)	0	2005–2018
Type 45	6,3 (9 total)	0	2004–2012
MARS	0,10	0	2007–2020
FSC/Future Minehunter	0,14/4	0	2007–2023
Future Submarine	0,1	+1	2009–2023

---

[5] We estimated this date to be when the *Vanguard* class would begin to come out of service.

[6] Because the only modification in this scenario is a Future Submarine, the baseline demand mirrors that of the full demand of this second scenario, minus the Future Submarine.

**Figure 2.12
Scenario 2: Addition of Future Submarine—Labour Projections by Programme, 2004–2025**

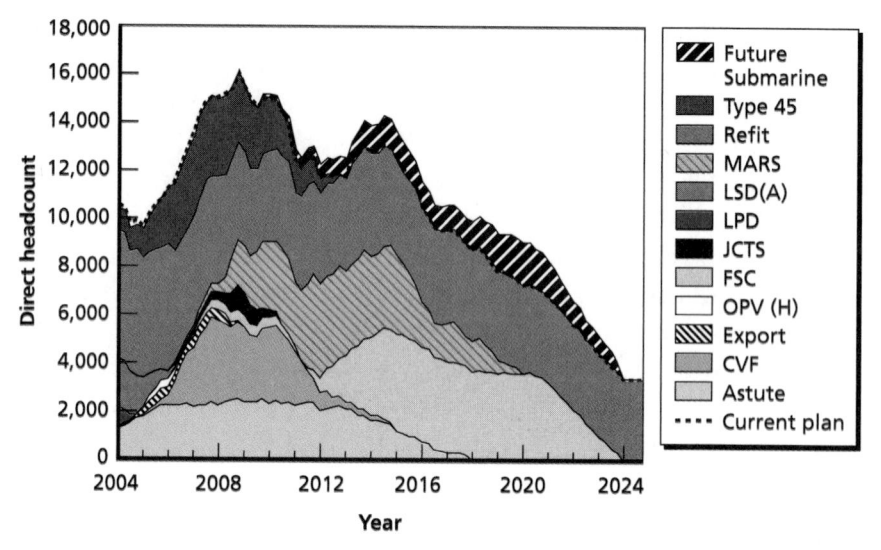

However, the most interesting impacts of adding a Future Submarine to the industrial base can be seen in the impact on our skill trade categories. Figure 2.13 plots these future demands for the skill trades.

Although the outfitting and structural trades show slight demand increases to account for the addition of a Future Submarine in the MOD's plans, the real difference can be seen in the technical skills. Designing a new submarine is a technically demanding activity, and we estimate that the industrial base would see an increase in demand from the MOD on the order of 1,000 technical workers from 2012 to 2015. This technical demand comes on top of a similar demand increase from 2005 to 2007–2008.

Figure 2.13
Scenario 2: Addition of Future Submarine—Labour Projections by Skill Trade, 2004–2025

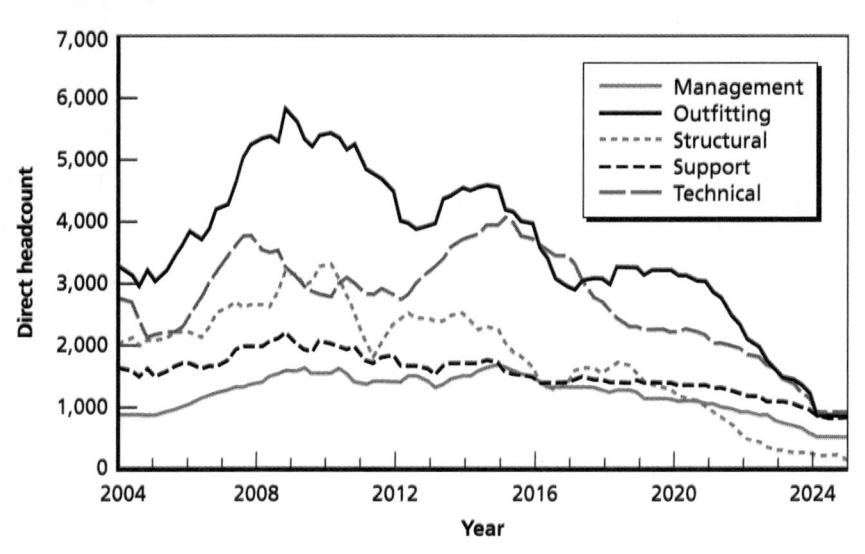

[Chart showing direct headcount from 2004 to 2024 with lines for Management, Outfitting, Structural, Support, and Technical]

RAND MG294-2.13

Scenario 3: Increased MOD Future Requirements

In our final scenario, we considered the impact on the industrial base if the MOD were to increase its operational requirements beyond that of our earlier assumptions. This situation would practically result in the MOD ordering more ships and, thus, placing an increased demand on the industrial base. The programme assumptions we used for this scenario are shown in Table 2.5 with the shaded sections showing the changes from the current MOD plan.

As the table shows, we increased the size of many of the programmes. We added two additional OPVs that will be built in the 2012–2014 time frame. We added two additional Astute submarines

Table 2.5
Scenario 3: Increased MOD Future Requirements—Programme Assumptions

Programme	Class Size (on contract, additional)	Delta Number of Ships from 'Current Plan'	Design/Build Duration (from 2004)
OPV(H)	2,2	0	2004–2006
JCTS	0,1	0	2006–2010
CVF	0,2	0	2004–2015
Astute	3,6 (9 total)	+2	2005–2020
Type 45	6,4 (10 total)	+1	2004–2013
MARS	0,10	0	2007–2020
FSC/Future Minehunter	0,18/6	+6	2007–2026
Future Submarine	0,1	+1	2009–2023

and one more Type 45 destroyer. The number of FSCs increased to 18 with its similar-hulled Future Minehunter also increasing from four to six. Finally, we kept the Future Submarine in this scenario. In addition to increasing the number of ships to be built, we adjusted the design/build duration periods to take these increases into account.

As one would expect, if the MOD were to increase its future requirements, the amount of demand on the overall industrial base would also increase. Figure 2.14 shows this impact.

Although future demand increases in later years, the changes made do not affect demand over the next six to seven years. Thus, the peak demand placed by this scenario is identical to the current MOD plan. The impact of these increased requirements begins in 2011, when future demand decreases at a much slower rate than in earlier scenarios. The same trends can be seen when looking at general skill levels, as evidenced by Figure 2.15.

Initially, the general skill trade demands act much the same as they did under the current MOD plan. However, later in time, all the skill trades show a marked slowing in their rate of decrease. As with scenario 2 (addition of Future Submarine), there is still an oscillating technical skill demand over time, which may pose a challenge for the MOD.

Labour Demand 49

**Figure 2.14
Scenario 3: Increased MOD Future Requirements—
Labour Projections by Programme, 2004–2025**

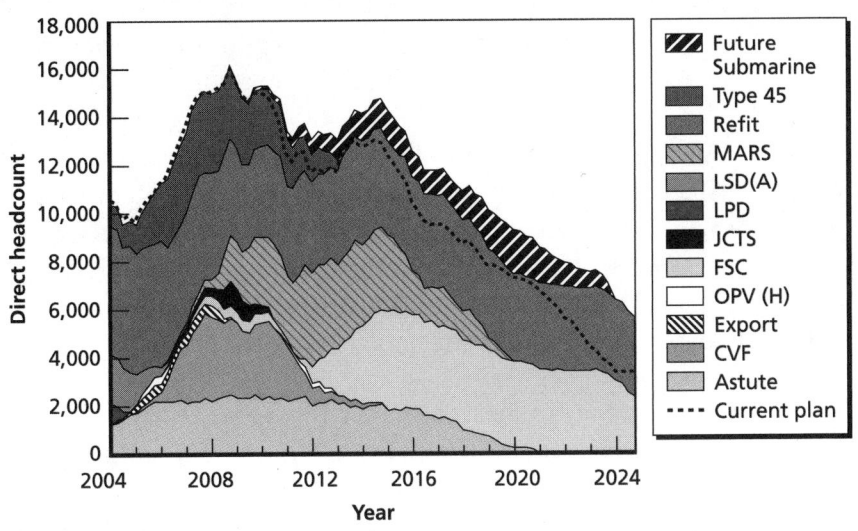

**Figure 2.15
Scenario 3: Increased MOD Future Requirements—
Labour Projections by Skill Level, 2004–2025**

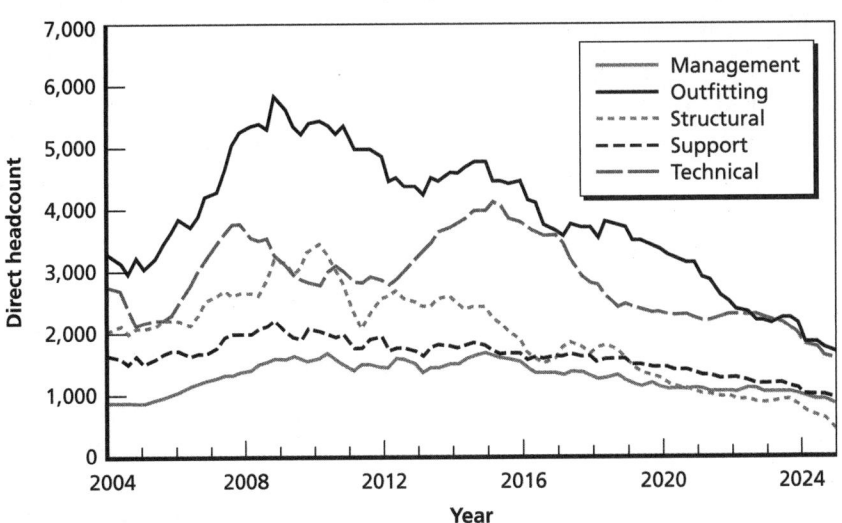

Looking at all three alternate scenarios, a few trends stand out. First, regardless of the level of future requirements, the MOD's demand on the industrial base over the next seven or eight years will be substantial, and it is in this period that many of the current challenges lie. Second, should the MOD add an additional Future Submarine to its requirements, it will need to carefully consider the impacts, especially on technical skills, of this choice. Finally, in all the scenarios, demand generally tapers into the future. This tapering is greater or less, depending on the scenario. Although it is difficult to make projections 15 to 20 years away, it does not appear that there will be substantial MOD programmes during this period—and that may have some long-term impacts on the industrial base.

Future MOD Programme Challenges

After closely examining the future MOD naval shipbuilding programme, both in the assumed current MOD plan and some possible variants of it, it is clear that the programme raises several challenges for the MOD and wider UK shipbuilding industrial base.

First, regardless of programme variation, the MOD's plans call for a dramatic increase in the amount of labour required to build its future ships. This increase in labour demand will force the shipbuilding industrial base to rapidly increase its workforce, especially in specific outfitting, structural, and technical skills.

Second, after the period of peak labour demand, the amount of direct workers needed to build the future MOD ships will decrease and will continue to decrease into the foreseeable future. This decline of workers may raise concerns about the long-term stability of the shipbuilding industrial base.

Third, the MOD's future shipbuilding programme creates specific challenges in managing the technical skills in the industrial base. With a short-term decrease in demand followed by a rapidly increasing need for technical skills, the industrial base may face significant challenges in either retaining or recruiting technical workers to meet the MOD's requirements.

Finally, although not discussed specifically in this chapter because of the commercial sensitivity issues, there may be workforce labour challenges in individual shipyards due to misalignment of skills, lack of future work, or periods of peak demand followed by periods of relative inactivity. These challenges are shipyard specific, but the MOD needs to be aware of them so that it can better manage its programmes.

These issues, though not insurmountable, will challenge the MOD and industry over the short-to-mid term. However, in the next section, we discuss possible ways to smooth the future labour demand to mitigate these challenges.

Options for Managing Increased MOD Demand

As discussed above, the future MOD shipbuilding programme will place increased demands on the UK shipbuilding industrial base. In this section, we examine steps that the MOD can take to help manage this increased demand to make its programmes more achievable.

'Level-loading' is one way the MOD could make it easier for the industrial base to meet future labour requirements. Level-loading is a general term used to describe a number of options that the MOD can use to smooth projected labour demand over time. Instead of continuing to follow the historic MOD demand profile of peaks and troughs, level-loading aims to smooth the labour demand so that industry can better manage its own workforce and prepare for future projects. The main advantages of this method are that it is directly within the control of the MOD to implement and that levelling workload will tend to level funding as well.

The MOD can use two basic techniques to put level-loading into practice:

- extending the length of programmes
- moving programmes forwards or backwards in time.

Additionally, the MOD could employ two other options that, though not necessarily reducing the overall projected labour demand, will help to shave peak labour demands at either the overall industry or individual shipyard level. These options are

- executing modular build strategies for programmes
- utilising smaller shipyards or nontraditional builders to construct blocks during peak demand periods.

In terms of level-loading techniques, the MOD can first look to extending the length of programmes. It can do this either by stretching the build time for individual ship-builds or by extending the time between ship-builds. Both methods can lower peak demands (and fill in troughs), depending how they are used. We provide an example of the latter technique in Figure 2.16.

The figure shows that, by increasing the interval between builds for a current programme from six months to nine months, peak labour demand, as depicted by the circles, would be reduced by 19.8 percent and the overall programme extended by one year.

Figure 2.16
Level-Loading by Extending the Time Between Ship-Builds

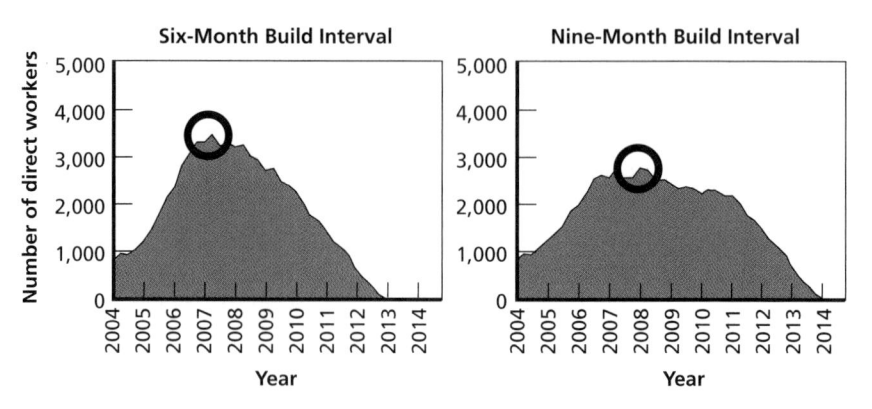

RAND *MG294-2.16*

Another level-loading technique that the MOD can use is to move programmes forwards and backwards in time. By shifting programmes away from periods of peak demand and re-programming them into periods of lower demand, the MOD can reduce the peak labour demands on the industrial base. Despite the level-loading benefits of doing so, we recognise that there may be challenges to this in practice. Moving programmes backwards in time (i.e., 'to the right') may not be possible because of operational restrictions. Existing programmes may be scheduled to come out of service at a specific time, and delaying programmes may create an unacceptable operational gap. Moving programmes forward in time (i.e., 'to the left') may also pose problems. Moving programmes forward without considering the impacts to design maturity may lead to cost overruns and further delays. This option should only be contemplated with careful consideration to design maturity issues.

Figure 2.17 shows an example of the potential benefits of moving a programme forwards or backwards in time.

Figure 2.17
Impact of Moving Programmes to Avoid Peak Demand

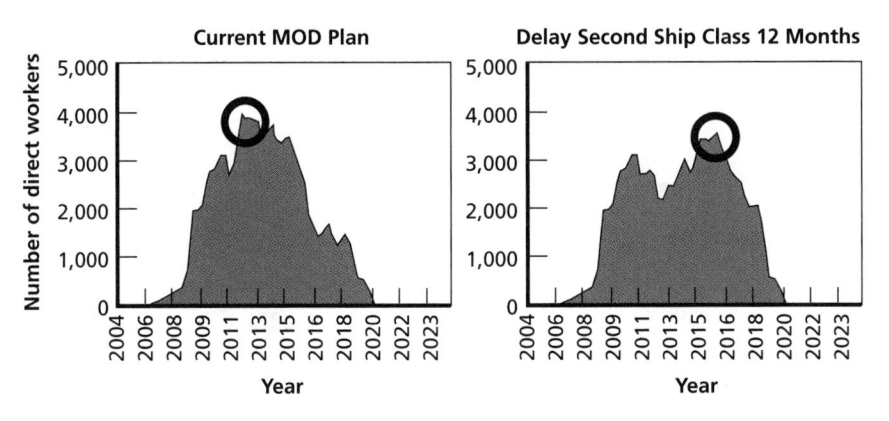

RAND MG294-2.17

In the figure, we looked at the impact of moving one of the ship classes of a future MOD programme to the right by 12 months. By moving the entire ship class to the right, the peak labour demand, identified by the circled areas, would be reduced by 11 percent, and the peak is delayed by four years. Moreover, the overall programme length remains the same.

Of course, it is possible to combine these two techniques and lower peak demand even further. Figure 2.18 shows the results of both moving the second class of a future MOD ship programme to the right by 12 months and extending the entire programme by 12 months.

The figure illustrates that by using two of the level-loading techniques—extending programme length and moving programmes around—it is possible to reduce peak demand. In this example, these techniques would reduce peak demand by 13 percent and would delay peak demand by five years. However, as we cautioned earlier, this may not always be feasible because of fiscal constraints, design

Figure 2.18
Impact of Both Extending and Moving Programmes to Avoid Peak Demand

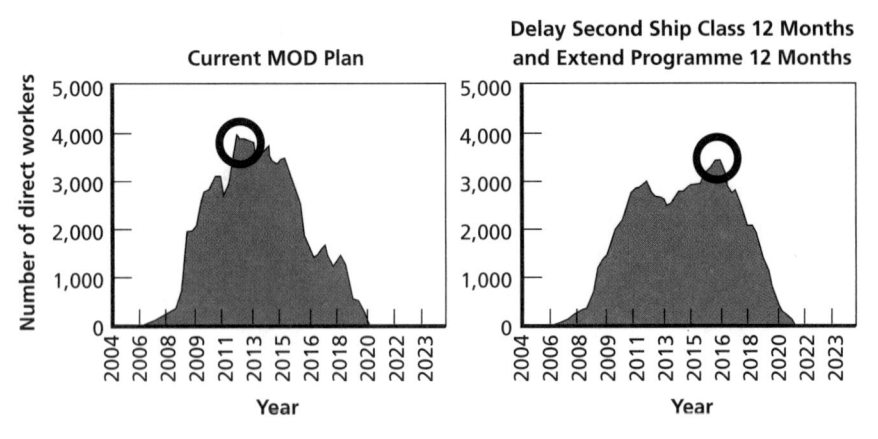

maturity issues, or replacement and capability needs. All these examples should be seen only as illustrative and not as a specific recommendation for the MOD.

Illustrative Results of Level-Loading Future MOD Labour Demand

Using the techniques defined above, we applied a series of changes to our original assumptions to see whether it is possible to 'level' the projected future labour demand the MOD will place on the industrial base through its naval shipbuilding and repair programmes. These changes, which are illustrative only, entailed the following:

- extending the interval between Type 45 builds to nine months
- moving JCTS forward by 18 months
- delaying one MARS ship class by 12 months
- extending all MARS ship classes by 12 months
- delaying the CVF programme by 12 months.

Inputting these changes into our labour estimation model allowed us to generate a new labour demand estimate (shown in Figure 2.19).

Figure 2.19 shows that, by making the five level-loading programme changes, the MOD could reduce its peak demand by approximately 12.5 percent[7] and delay the date of peak labour demand by just under three years. The labour projection graph is also noticeably flatter and reduces the slope of the labour decline after the CVF and Type 45 programmes finish.

Another way to graphically see how this level-loading scenario smoothes demand is to look at the change in labour demand over time compared with the current demand (as of 2004) placed by the MOD (Figure 2.20).

---

[7] This percentage corresponds to roughly 1,900 direct workers.

Figure 2.19
Level-Loading Labour Projections, by Programme

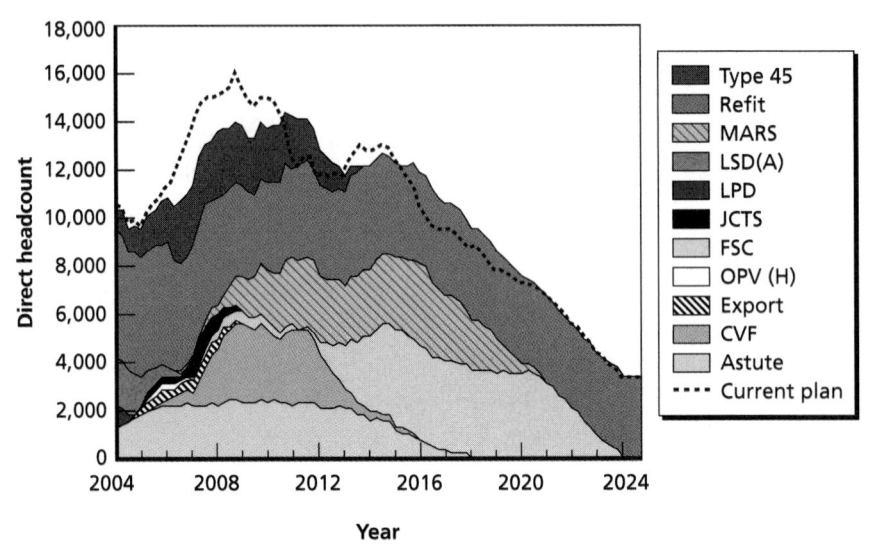

RAND MG294-2.19

Figure 2.20
Base Case and Level-Loaded Demands Compared with Current MOD Demand

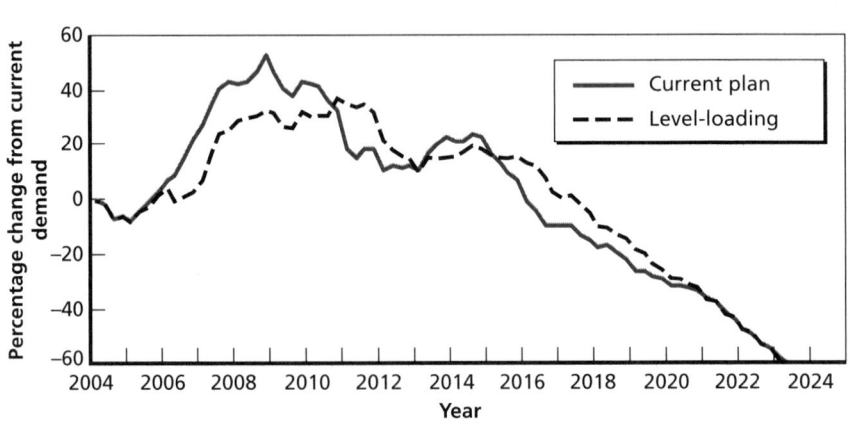

RAND MG294-2.20

As Figure 2.20 shows, under the current MOD plan, the peak demand would rise to more than 50 percent of the current MOD demand, while the level-loaded case would rise only about 35 percent above the current demand. The figure also shows that the changes in demand also occur more gradually and should be easier for the industrial base to manage.

Applying these level-loading assumptions to the general skill trades, we also find that level-loading improves the demand profile for almost all trades (see Figure 2.21).

Under level-loading assumptions, the peak labour demand would decline for all skill trade categories, although the decrease in technical demand in 2005 would still be followed by a rapid upsurge in demand. Regardless of level-loading strategies used, we believe this short-term challenge will remain.

Figure 2.21
Level-Loading Labour Projections, by Skill Level

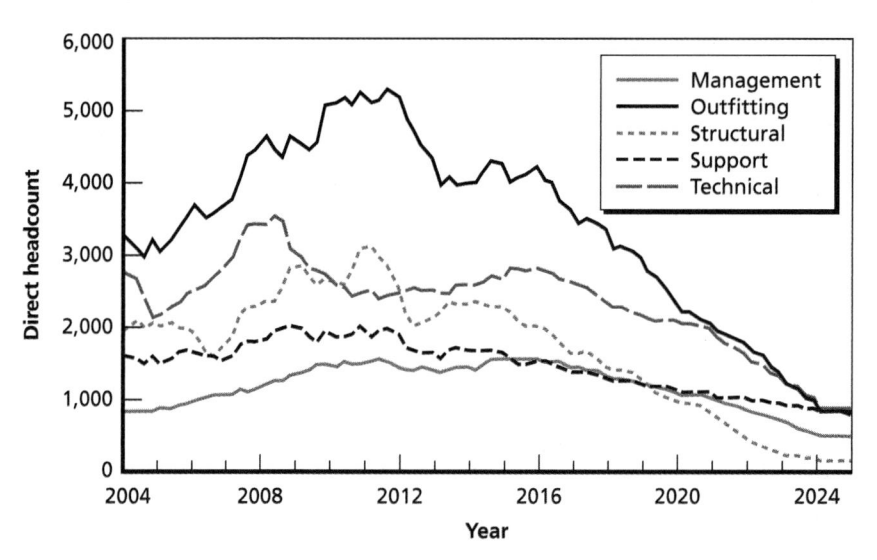

As we mentioned earlier, these figures are meant to be illustrative to show *how* the MOD may go about levelling its future demand. We do not claim that the level-loading changes we made to the current MOD plan are optimal or even feasible. They serve only to illustrate ways in which the MOD could reduce peak demand. Specific decisions that the MOD could make to reduce peak demand should be taken only after considering their impact on the totality of its shipbuilding programme and how they may affect existing programmes and the industrial base as a whole. These specific recommendations will require further analysis by the MOD and are beyond the scope of this study. Most importantly, the MOD will need to balance the industrial needs with its operational requirements in any levelling plan.

Other Build Strategies

In addition to employing level-loading techniques to smooth out periods of peak demand, the MOD can shave individual shipyard peaks by employing two additional strategies.

First, the MOD can utilise modular build strategies similar to those used on the Type 45 programme. By having both BAE Systems and VT Shipbuilding build blocks for the Type 45, labour is distributed between both shipyards and reduces the peak demand at any one shipyard. In terms of reducing peak demand at an individual shipyard, this strategy would work well, especially for larger ships such as the CVF and MARS.

Second, the MOD can encourage its programmes to use smaller and medium-sized shipbuilders or nontraditional marine firms during periods of peak demand. This use will reduce the load on the larger, core shipyards while providing work for smaller firms that may not have the capability to build large ships on their own. There are a variety of shipyards in the United Kingdom that have the potential to fulfil this role, including Appledore, KBR, AMEC, Harland and Wolff, Ferguson, the A&P Group, Northwestern Shiprepairers, and the former Royal Dockyards at Devonport, Portsmouth, and

Rosyth.[8] This list is not exhaustive but does illustrate the breadth of additional resources that may be potentially available to the United Kingdom. However, before committing to use additional firms in this way, the MOD must ensure that these shipyards have the capability (in both facilities and skills) to fulfil these 'peak-shaving' roles.

The MOD can also consider the fabrication and project management skills available in the offshore industry. We specifically address the capacity of this industry, which may have a part to play in helping the MOD to complete its future programmes, in Chapter Six.

Summary

In this chapter, we have looked at the future MOD shipbuilding programme (as we see it in the next 20 years) and have analysed the labour demands that it will place on the UK shipbuilding industrial base. We found that, over the next six years, demand will increase significantly, after which point it will decrease to below current levels. The rapid increase in demand, largely caused by four major programmes—Astute, Type 45, CVF, and MARS—will force the naval shipbuilding industrial base to appreciably increase its workforce.

Moreover, there are significant differences between specific labour skills. The manufacturing trades (i.e., outfitting, structural, and support) show increases from demand levels in 2004 of between 33 and 81 percent—the largest increase being outfitting. The management category follows a similar trend but grows by nearly 90 percent over the 2004 demand level. The trend for technical skills is more complicated. Initially, there is a short-term drop (2005) in demand for technical workers—a drop of approximately 22 percent relative to 2004. This decrease is followed by a rapid increase in demand to a level about 40 percent *greater* than the 2004 level.

---

[8] In fact, many of these firms have a record of producing either warships or support ships for the Royal Navy or Royal Fleet Auxiliary. However, this experience lies largely in designing and producing smaller or less-complex vessels.

Clearly, managing the technical workforce over these rapid changes will be problematic.

By employing a variety of level-loading techniques, the MOD can mitigate the labour challenges posed by its planned shipbuilding programme and both decrease and delay the peak labour requirements on the industrial base. The MOD will have to carefully consider how to do this, as it needs to balance its operational and fiscal requirements with industrial base concerns.

Another way to reduce peak demand periods, especially at individual shipyards, is to utilise smaller shipyards or portions of the wider marine industry in parts of the future shipbuilding programme. Again, the MOD will have to consider how to do this on an individual basis, since different firms have their own core competencies.

CHAPTER THREE
The Supply of Naval Shipyard Labour in the United Kingdom

Whereas the preceding chapter discussed the demand for specific shipbuilding skills in the United Kingdom, this chapter addresses the supply of skills in the shipbuilding and repair industry.[1] The major focus here is whether and how the shipbuilding workforce in the United Kingdom can expand to meet the upcoming surge in naval production—a difficult analytical issue. First, the data on the shipbuilding workforce in the public domain are inconsistent and incomplete, despite an ongoing effort to build more comprehensive workforce databases, particularly at the regional level. Second, many factors that might help resolve the labour problem are uncertain and difficult to quantify. That said, we have collected data from the major British naval shipbuilders and repairers, shipbuilding-related companies, regional development agencies, and private personnel agencies that allow us to generally forecast the potential pool of workers within the UK shipbuilding industry from 2004 to 2020.

---

[1] In this chapter, we have collapsed the range of required shipyard skills into two major categories of workers: management/technical and manufacturing. (Management and technical workers are employees involved in administration, marketing, design, engineering, estimating, and project management, among other skills. The manufacturing workers are those who perform structural, outfitting, and direct support tasks.) We have done this because not all the UK firms that we contacted for data on the potential supply of naval shipyard labour could easily provide information on worker recruitment and attrition at the subcategory level, so making supply projections at that level would seem to imply a level of precision that we could not justify. Nevertheless, this chapter will present qualitative, if not quantitative, information on the difficulties that the shipyards have been experiencing in specific skill areas.

In this chapter, we begin by reviewing what is known about the employment situation in the UK shipbuilding industry. We then describe some of the factors that may constrain or enhance the ability of the naval shipyards to expand their workforce. Finally, we address the question of how the potential supply of naval shipyard workers compares with the demand for workers under different supply and demand scenarios.

Employment Status of the UK Shipbuilding and Repair Industrial Base

Until recently, the number of persons directly employed in the UK shipbuilding, repair, and offshore sector had been declining for decades. According to the Office of National Statistics (ONS), industry employment in Great Britain (not including Northern Ireland) stood at almost 41,600 in 1991 and about 25,000 in 2000, where it has more or less remained (see Figure 3.1).[2]

Regional Differences in UK Shipyard-Related Employment

Despite the overall decline in shipyard related employment in Britain, regional employment in this industry has varied. According to the ONS, there was a relatively consistent downward trend in employment in Scotland (from 11,516 workers in 1995 to 8,386 workers in 2000), the northwest (from 7,758 workers in 1993 to 1,652 workers in 2000), and the northeast (from 8,078 workers 1996 to 1,633 in 2000). However, the trend was less clear in the southwest and southeast regions of the country. Employment edged up in the southwest from 5,800 workers in 1998 to 6,471 workers in 2000 and remained

---

[2] The ONS data were sorted by Standard Industrial Classification 35.11 (Building and Repairing of Ships) and include full-time and part-time workers involved in the building and repairing of commercial vessels and warships, as well as in the construction of offshore drilling platforms and floating structures.

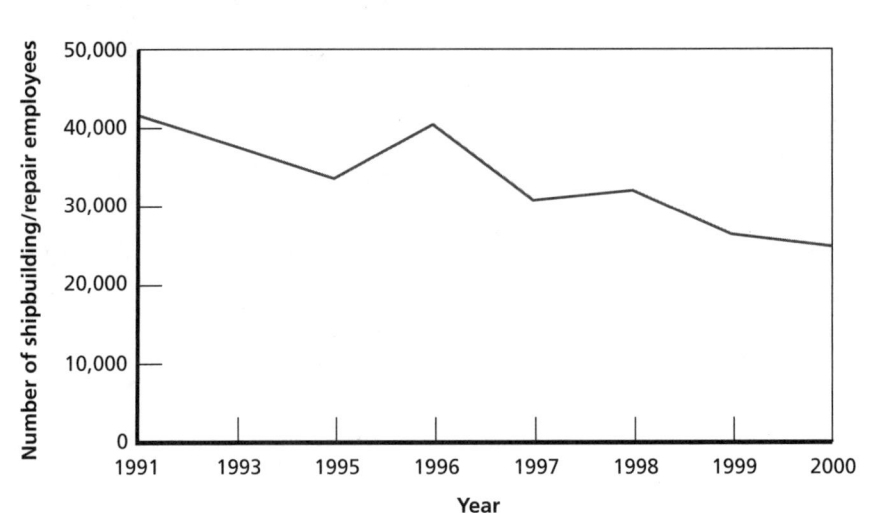

Figure 3.1
UK Shipbuilding, Repair, and Offshore Employment in the 1990s

[Line chart: Number of shipbuilding/repair employees vs. Year (1991–2000). Values start around 41,000 in 1991, decline to ~33,000 in 1993, rise to ~40,000 in 1996, then decline to ~25,000 by 2000.]

RAND MG294-3.1

fairly stable in the southeast in the 1990s at between 4,000 and 6,000 workers. Figure 3.2 depicts the plotting of these regional trends. It should be noted that the ONS data on shipyard employment for northwest England in 2000 are in substantial conflict with information supplied by BAE Systems on employment at its Barrow facility in 2000, which shows higher levels of employment in the northwest.

Sector Employment in the UK Shipbuilding and Repair Industry

Although our sources generally agree on the total number of employees in shipbuilding and related industries, they are less clear about how many of these people work in the separate categories of merchant shipbuilding, naval shipbuilding, ship repair, and offshore oil fabrication. In 2000, fewer than a quarter of those employed in the overall shipbuilding and repair sector worked in commercial ship-

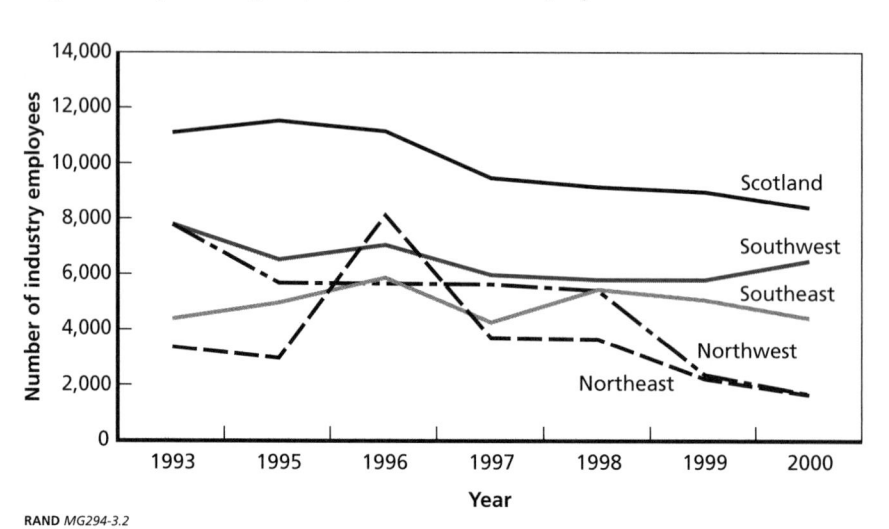

Figure 3.2
Regional Shipbuilding, Repair, and Offshore Employment in the 1990s

building and ship repair, as well as offshore oil construction.[3] As Figure 3.3 indicates, more than three-quarters of those in the shipbuilding and repair sector worked as builders or repairers of naval vessels.[4]

According to the responses to the questionnaire that we sent to UK naval yards in fall 2003[5], the number of manufacturing employees in the industry increased from 9,096 in 1999 to 10,425 in 2001 before falling back to 9,685 in 2003 (see Figure 3.4). This rise is

---

[3] The available data do not distinguish between employment in the offshore and commercial shipbuilding/repair industries.

[4] This picture of the numbers of workers employed in various sectors of the shipbuilding and repair industry was pieced together from a number of sources, including the ONS, Appledore/University of Newcastle (*Prospects for UK Merchant Shipbuilding Industry*, 2000), and the Confederation of Shipbuilding and Engineering Union's (CSEU's) Shipbuilding Working Group.

[5] These naval shipyards include those owned by BAE Systems in Barrow-in-Furness and Glasgow, VT, FSL in Portsmouth, DML in Plymouth, Swan Hunter in Newcastle-upon-Tyne, and BES in Rosyth.

The Supply of Naval Shipyard Labour in the United Kingdom   65

Figure 3.3
Share of Workers in Shipbuilding and Repair Subsectors in 2000

- Commercial shipbuilding/repair and offshore construction (24%)
- Naval ship repair (40%)
- Naval shipbuilding (36%)

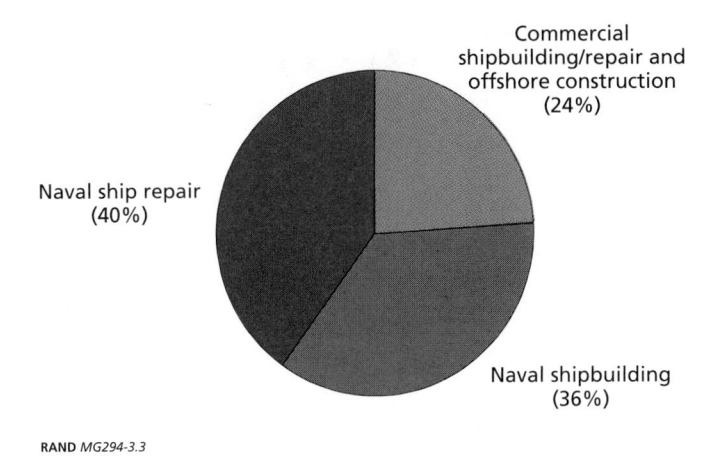

RAND MG294-3.3

Figure 3.4
Number of Workers in the Naval Yards, 1999–2003

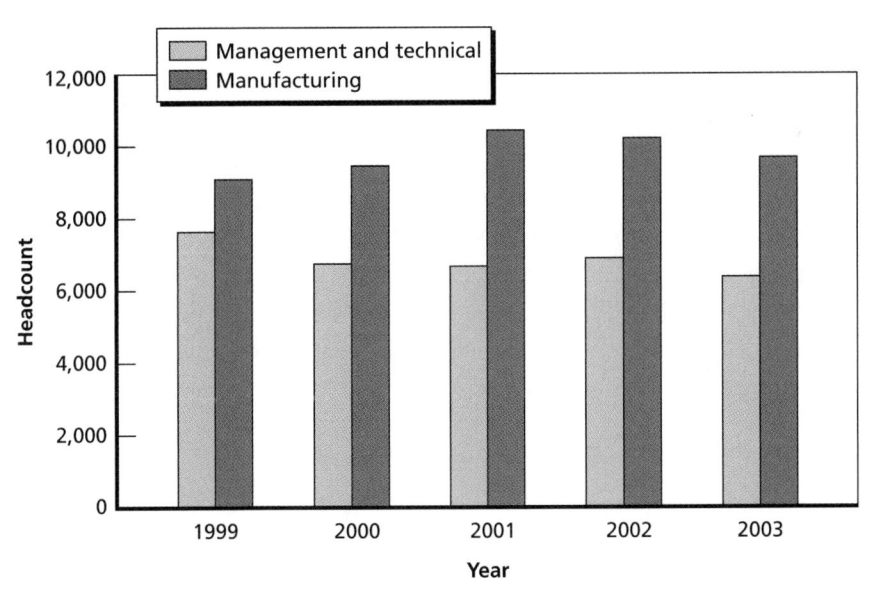

RAND MG294-3.4

largely attributed to the reopening of Swan Hunter and an increased repair/refit workload at DML's Devonport Royal Dockyard in Plymouth. Following a drop-off in 1999, the number of management and technical workers in the seven naval shipyards has remained relatively steady: It stood at 6,396 in 2003. The only significant increases in recent years occurred at DML and Swan Hunter.

UK Shipbuilding and Repair Industry Workers Are Ageing
Despite the relatively stable rate of employment in naval shipbuilding and repair industry in recent years, the workforce is ageing. According to our 2003 survey, almost 60 percent of the management and technical workforce in the seven naval yards is more than 40 years old (see Figure 3.5). One-third of the managers and technicians in the six yards are between the ages of 41 and 50. About a quarter are 51 to 60 years old, and fewer than 3 percent are over 60 years old. Only 14 percent of managerial technical workers are 30 years old or younger —although BAE's Glasgow shipyards and Swan Hunter have a more youthful component of management and technical employees than the other yards. The important 31–40 age group comprises 26 percent of the total management and technical workforce, but its size varies considerably across the shipyards, from a low of 16 percent to a high of 33 percent.

The manufacturing workforce in the naval shipbuilding and repair industry is even older than the management and technical workforce. Sixty-two percent of manufacturing employees in the six core shipyards are older than age 40. About a fifth of manufacturing workers are in the 31–40 age group. Although 17 percent of all manufacturing workers are 30 years or younger, the size of this group varies considerably, from 5 to 25 percent of a yard's manufacturing workforce.

The flip side of the ageing workforce story is that naval shipbuilding and repair workers in the United Kingdom are generally quite experienced in their jobs. More than 75 percent of the management, technical, and manufacturing workers have more than five years of experience. Only in one case does fewer than 60 percent of the shipyard's overall workforce have five or fewer years of experience.

Figure 3.5
Age Profile of the Workforce in the Naval Shipyards in 2003

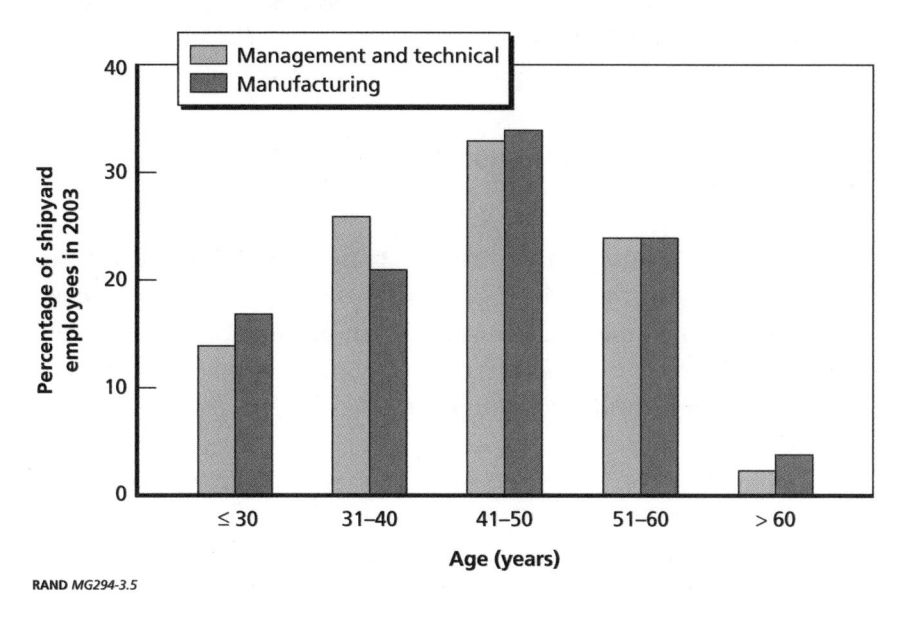

[Bar chart showing percentage of shipyard employees in 2003 by age group, with bars for Management and technical, and Manufacturing. Age groups: ≤30, 31–40, 41–50, 51–60, >60]

RAND MG294-3.5

Small Reliance on Temporary Workers

One way that the naval shipyards could grow to meet any potential surge in demand during the next several years is to hire more temporary workers. However, with the exception of BES Rosyth and, to a lesser extent, BAE Barrow, most naval shipyards do not currently use many (if any) temporary workers. Two shipyards employed no recruitment agency personnel in 2003.[6] According to officials at one shipbuilding company, they have more than adequate personnel resources in their trades' database and claim that the use of temporary workers discourages the formation of a stable, permanent labour force. Another shipbuilder does not currently employ many temporary workers but anticipates it may in the future. In addition, the same shipbuilder recently began recruiting significant numbers of

_____

[6] It should be noted that the demand for naval vessels was relatively low in 2003. Some shipyards may have used more temporary workers during peak periods in the past.

staff on short- or fixed-term contracts to enlarge the number of people in its region with shipbuilding experience, should they be required to meet future demand.

Ability of the Naval Shipyards to Expand Their Workforces

Recent studies provide an uncertain picture of shipbuilding employment.[7] On the positive side, they all agree that the current workforce is large enough to meet current naval ship production requirements. They also note that the shipyards are confident in their ability to rapidly expand their workforces to satisfy increased workload requirements.

Such expansions could occur in most regions. In Scotland, shipbuilders expect to take on an additional 2,300 employees. Swan Hunter reports that the size and flexibility of its available workforce has meant that ramping up resources for large projects has rarely caused a problem, as exemplified by the company's recruiting of 3,000 workers in about six months for the Solitaire pipe layer project in the late 1990s. For its part, VT Shipbuilding has said that it could cope with the demand for labour under any potential scenario through a combination of recruitment, seconding of workers from FSL, and the use of temporary employees. Notwithstanding its location in southwest England, which is geographically removed from the employment hub, DML has doubled its workforce at times, if only for a short period.[8]

Concerns About Labour Shortages

Shipyard sources have expressed concern about the workload gap between 2003 and 2006, during which time shipyard owners may lay off workers they may need in the future. In particular, the shipyards

---

[7] See, for example, Birkler et al. (2002) and Bruce (2002).

[8] This doubling of the workforce at DML occurred in the case of the D154 nuclear facilities upgrade.

are worried that, unless the MOD starts other programmes (such as MARS) earlier than planned, shortages of certain kinds of highly skilled workers, such as design engineers for CVF and later programmes, might emerge. Once made redundant, they believe, many of these highly skilled persons will not return to the shipbuilding profession.

Shipyard officials make a distinction between workers with generic skills and those with domain expertise. Although the former may be well educated (e.g., finance or human resource specialists), they do not represent a critical resource. Those with domain knowledge—defined as an intimate knowledge of naval shipbuilding rules and standards—are key to a company's naval business. These people often take a long time to train. For example, it can take test and commissioning engineers 10 to 20 years working in the industry to become fully proficient. Furthermore, these engineers cannot be easily replaced in the short term by technical experts from other industries or even other shipbuilding fields (e.g., submarines or naval surface ships). Finally, there is a limited pool of suitably qualified people with necessary domain experience who are not already employed in shipbuilding.

Recruitment in the Shipbuilding and Repair Industry Faces Significant Obstacles

Despite the shipyards' basic optimism regarding their recruitment capabilities, recruiting and retaining workers in the shipbuilding industry is becoming more difficult. Educated young people have more job opportunities than they have had in the past, and they are less likely to take positions in industries, such as shipbuilding, that are perceived as being strenuous, uncomfortable, and unstable. For example, a survey of 1,063 second- and third-year undergraduates in the shipbuilding community of Barrow-in-Furness indicated that only 115 are undertaking studies likely to be of relevance to shipbuilding.[9] Furthermore, those with skills of great importance to shipbuilding, such as electricians, have good prospects in other industries.

The recruitment problem is exacerbated by the decline in the shipbuilding training infrastructure. According to the Engineering Training Maritime Authority, most shipbuilding and repair companies are ineligible to receive public support for training because of their large size. However, small companies that are eligible for government funding are reluctant to give up productive employees for training or mentoring.[10] Shipyard officials told us that the current apprentice training programme does not motivate people to work in trades such as shipbuilding. The National Vocational Qualification programme, they say, has significant academic requirements that are difficult for the average tradesperson to fulfil. Those who do meet these requirements often go on to obtain a full college degree, and then expect desk jobs rather than waterfront positions.

Shipyard Training Initiatives

Still, significant steps are being taken to remedy the recruitment problem. According to the information provided to us in fall 2003, most of the major shipbuilders have teamed with local community development agencies and schools to expand apprentice programmes and entice both younger and older workers to enter the shipbuilding and repair profession.

In Scotland, Scottish Enterprise Glasgow has helped implement skills training for BAE Systems as part of the Clyde Shipyards Task Force. In 2003, it established a Construction Skills team with resources of £25 million, which may soon be available to the shipbuilding sector. Scottish Enterprise also has plans for a Scottish Marine Technologies Training Project that would provide on-site accreditation and adult training on the job. In addition, BES and

---

[9] Furness Enterprise Limited (2003). According to Scottish Enterprise, Careers Scotland is attempting to improve the image of manufacturing professions such as shipbuilding. For example, it has targeted its 'Make It in Scotland' campaign at 11- and 12-year-olds before they make their exam subject choices.

[10] Closhen (2002).

Lauder College are developing improved procedures for selecting and training craft modern apprentices, as well as programmes for up-skilling, re-skilling, and cross-skilling existing employees.

Swan Hunter's training programmes cover all disciplines and include modern apprentice schemes, adult training and retraining schemes, graduate training and development, management training and general re-skilling. Swan Hunter and its business partners plan to at least double their current apprentice intake over the next 10 years. The company is piloting a fast-track apprenticeship programme for adults 25 to 40 years old, who must complete the same course as modern apprentices but do so within two years as apposed to the current four-year modern apprentice scheme for trainees aged 17 to 24.

The Northwest Development Agency has recently awarded a training grant to BAE Barrow.

In southern England, VT Shipbuilding has joined with a number of organisations—for example, the Learning and Skills Council, Engineering Employers' Federation, and the SEMTA—to promote careers in engineering, provide vocational training for young people and older adults, and establish a set of specific occupational standards. VT is also working with Nottingham Trent and Portsmouth universities to develop a modular, or points-based, vocational degree that could be delivered in the workplace as part of the 'Skills for Life' project.

Consequences of Unemployment, Demographic Changes, and Shipyard Redundancies

Many assume that a situation of high unemployment is good for shipbuilding recruitment. If true, the situation in the United Kingdom offers a mixed picture. On the one hand, the number of unemployment claimants has declined in all major UK shipbuilding and repair regions in recent years.[11] On the other hand, unemployment

---

[11] Between 1998 and 2001, the number of unemployment claimants declined in Scotland (from 140,000 to 107,000), Northeast England (from 83,000 to 60,000), Northwest England (from 166,000 to 127,000), Southeast England (from 106,000 to 66,000), and Southwest England (from 83,000 to 53,000).

levels in the northern communities of Glasgow, Dunfermline, Tyneside, and Barrow are relatively high compared with the southern shipbuilding communities of Plymouth and Portsmouth. For example, unemployment claimants as a percentage of workforce jobs were two-and-a-half times higher in Dunfermline, the home of BES, than in Portsmouth, the home of VT and FSL (see Figure 3.6). Thus, it is likely that northern shipyards will have a larger pool of unemployed workers from which to draw during periods of high demand than will southern shipyards.

It is also argued that the increase in the number of young people entering the workforce could help alleviate projected shipyard labour shortages. For example, in Scotland, the population of people at school-leaving age (aged 16 to 19) is projected to peak in 2004 at 264,000 and will not slip below 2002 levels (254,000) until 2011. The population of young adults (aged 15 to 29) will peak at 964,000

Figure 3.6
Unemployment Levels in Important Shipyard Towns in 2002

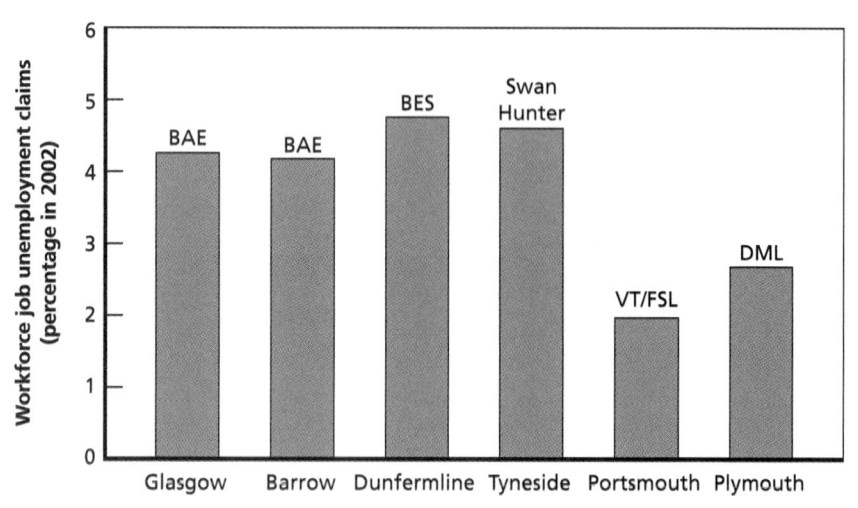

in 2008 and 2009, precisely when the shipyards will be most actively recruiting.[12]

Recent shipyard redundancies could also expand the size of the available labour pool in certain areas. The number of redundancies in the naval shipyards from 2001 to 2003 averaged 225 per year for management and technical workers and 720 for manufacturing workers. Many of these redundancies occurred at BAE's shipyards in Barrow and, to a lesser extent, Glasgow. Although older unemployed workers may provide a reservoir of workers during the projected surge in shipbuilding construction later in the decade, shipbuilding officials have voiced concern about rehiring them because their preexisting health problems could lead to lower productivity. In addition, research on previous downsizings at Barrow has shown that many redundant workers had taken other jobs, retired, or accepted incapacity or unemployment benefits. Furthermore, many of these ex-shipyard workers—including those who had found positions elsewhere as well as those who had left the labour market (the so-called hidden unemployed)—were unlikely to return to their former shipbuilding jobs in Barrow without a significant financial incentive.[13]

Recent Shipyard Recruiting Efforts

In general, recruiting in British naval shipyards has been rather stable in recent years. As Figure 3.7 shows, this is particularly true for the management and technical category of workers. An average of 390 recruits were hired per year into this category between 1999 and 2003. Among the shipyards, BAE Glasgow has been recruiting most heavily, while Swan Hunter and DML had a substantial surge in recruitment in 2001. BAE Barrow recruited a relatively large number of management and technical workers in 1999 and 2000, but the shipyard has since reduced its rate of hiring. In recent years, VT

---

[12] Private communication, official, Scottish Government's Actuary Department, January 2004.

[13] Furness Enterprise Limited (2003).

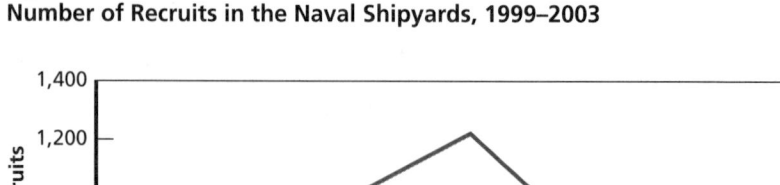

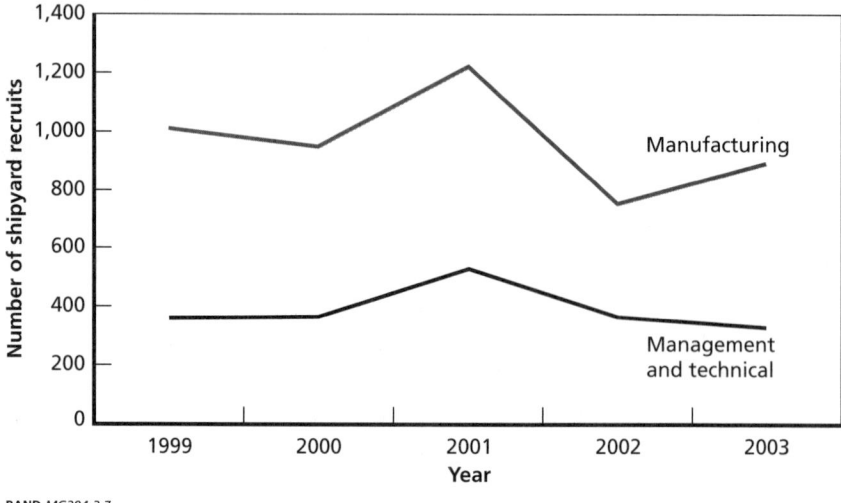

Figure 3.7
Number of Recruits in the Naval Shipyards, 1999–2003

Shipbuilding, FSL, and BES Rosyth have been recruiting management and technical workers at a relatively low level, although the latter two increased their hiring in this area in 2002–2003.

The number of manufacturing recruits averaged fewer than 1,000 annually during the 1999–2003 period. In 2001, however, it surged to more than 1,200 and then dropped below 750 in 2002 before rising to 892 in 2003. Much of this surge was attributable to Swan Hunter's recruiting following its reopening. Although BAE's Barrow shipyard significantly reduced the number of manufacturing recruits throughout this period, the company's Glasgow yards maintained a relatively consistent level of manufacturing recruitment. The same is largely true of DML. FSL recruited a substantial number of manufacturing workers in 2002. BES Rosyth and VT Shipbuilding have hired relatively few manufacturing workers since 1999.

Pools of Labour That Could Be Tapped

Previous shipbuilding capacity studies have indicated that there are a number of pools of labour that can be tapped by the naval shipyards to meet projected increases in workforce requirements between 2006 and 2018.[14] For example, a declining UK merchant shipbuilding industry offers a potential source of additional workers. Whereas UK companies held one-third of the world commercial shipbuilding market in 1950, it has only about a 1 percent share today. Not only has the United Kingdom had difficulty keeping pace with industry leaders Japan and South Korea, it has also lost ground to the Netherlands, Spain, and Germany because of the latter's lower prices and higher productivity.[15] Between 1990 and 2000, the number of ships delivered annually by British commercial shipyards fell from a high of 25 to five. Of the 28 yards that delivered ships during this period, only 12 remained active in 2000.[16]

Another possible source of shipyard workers is the offshore fabrication industry. After a continuous downturn in the 1980s and 1990s—which seemed to spell the end of the offshore industry in the United Kingdom—prices recovered in 1999, and the industry appeared to regain its sense of optimism.[17] Since then, however, offshore construction has remained stagnant, with operators reluctant to invest in new platforms, given the limited prospects for additional significant oil discoveries in the North Sea region. As a result, companies in the offshore business—such as KBR Caledonia, which has a large facility in Nigg, Scotland—are actively seeking shipbuilding production work. For their part, merchant shipbuilding, repair, and design companies are teaming with traditional naval shipyards to share the projected increase in the naval shipbuilding workload.

---

[14] In addition, Swan Hunter, Scottish Enterprise, and Highlands and Islands Enterprise have developed regional databases that contain thousands of workers with skills relevant to shipbuilding. However, it is unclear how many of these workers would be available to the shipbuilding industry in the future.

[15] Shipbuilders & Shiprepairers Association (2000).

[16] Appledore/University of Newcastle (2000, p. 21).

[17] See UK Offshore Operators Association (2000) and Bradbury (2000).

Although primarily involved in ship repair, BES and DML are undertaking shipbuilding tasks, and BAE Naval Ships has proposed sharing CVF work with ship repair yards and smaller ship-build facilities. In 2000, VT Shipbuilding reached an agreement with FSL management and its respective trade unions that enabled the two companies to temporarily transfer personnel. VT plans to use FSL's labour pool to cover the company's peak requirements for steelworkers, electricians, engineers/pipe workers, and those in the finishing trades.

Shipyards Rely on Outsourcing to Varying Degrees
Although outsourcing has increased substantially in the European commercial shipbuilding industry in recent years, UK naval shipbuilders have traditionally been unwilling to subcontract to any great extent, preferring to keep the bulk of the business to themselves. Their rationale has been that outsourcing has not worked well for sophisticated naval shipbuilding projects, pointing to problems that commercial shipbuilder Kvaerner experienced in the construction of the LPH, HMS *Ocean*. According to a recent RAND study, there are considerable differences in the amount of outsourcing that UK shipbuilders are undertaking.[18] As Figure 3.8 shows, Swan Hunter is the most active user of total outsourcing among the large shipyards. It works closely with NEMOC (North East Marine and Offshore Cluster), an alliance of complementary companies located and working in the Tyneside area. Tasks the company considers for subcontracting are painting, HVAC, electrical, and insulation installation. VT Shipbuilding is the most active user of peak

---

[18] Schank et al., (forthcoming). This study found that shipyards employ two types of outsourcing: total and peak. *Total outsourcing* involves a shipbuilder subcontracting a complete functional task, such as electrical, HVAC, or painting to an outside firm. In this case, the shipbuilder retains no in-house labour capability to perform the function, although it may provide facilities or material and equipment to the subcontractor. *Peak outsourcing* occurs when a shipbuilder uses a subcontractor or temporary labour to augment in-house capabilities during time of peak demand, to reduce the shipyard workforce when demands decrease if faced with strict national policies limiting the ability to terminate workers, or to accelerate operations when schedules start to slip.

Figure 3.8
Total and Peak Outsourcing Undertaken by UK Naval Shipbuilding and Repair Companies

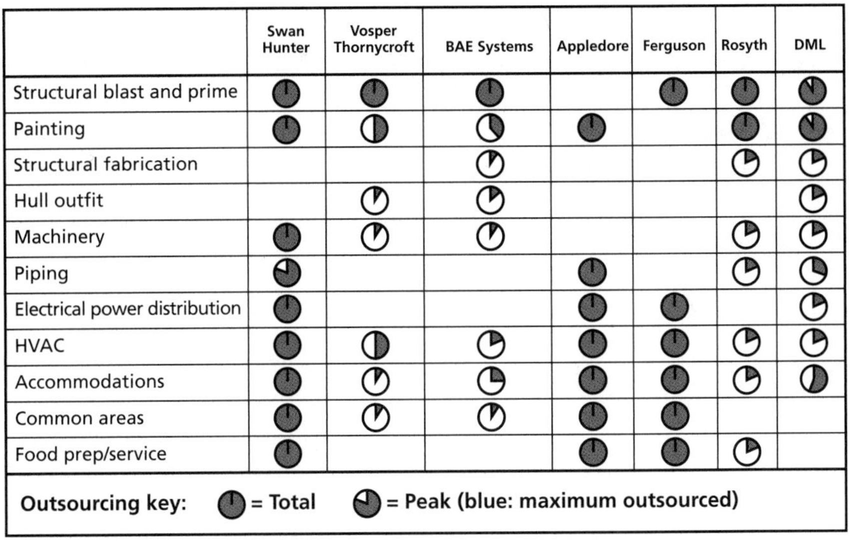

	Swan Hunter	Vosper Thornycroft	BAE Systems	Appledore	Ferguson	Rosyth	DML
Structural blast and prime	◐	◐	◐		◐	◐	◐
Painting	◐	◐	◐	◐		◐	◐
Structural fabrication			◐			◐	◐
Hull outfit		◐	◐				◐
Machinery	◐	◐	◐			◐	◐
Piping	◐			◐		◐	◐
Electrical power distribution	◐			◐	◐		◐
HVAC	◐	◐	◐	◐	◐	◐	◐
Accommodations	◐	◐	◐	◐	◐	◐	◐
Common areas	◐	◐	◐	◐	◐		
Food prep/service	◐			◐	◐	◐	

Outsourcing key: ◐ = Total   ◐ = Peak (blue: maximum outsourced)

RAND MG294-3.8

outsourcing. This southern English shipbuilder subcontracts 12 percent of total steelwork man-hours and 66 percent of total electrical man-hours. By contrast, BAE is the least active user of total or peak outsourcing, even though BAE's Govan site previously made extensive use of subcontractors when Kvaerner owned it.

A Comparison of the Supply of Naval Workers with the Demand Under Different Future Scenarios

Given the mix of constraints and opportunities faced by the shipyards in their ability to expand their workforce, how does the potential supply of shipyard workers compare with the demand under different supply and demand scenarios?

To answer this question, we built a simple spreadsheet model to try to forecast the labour supply for naval shipyards over the next 17 years. Because of the lack of comprehensive and consistent data, we did not attempt to optimise the labour force for individual shipyards. Instead, we focused on sources of attrition and recruitment identified in recent shipyard surveys and compared the total potential supply within the industry to the demand under the current MOD plan under several different scenarios.

Three Supply Cases

We examined three cases in our supply side analysis:

1. A 'no recruitment' case, in which the current direct workforce[19] in the United Kingdom's naval shipyards and other firms currently involved in government shipbuilding was reduced through retirement[20], long-term incapacity, and other 'voluntary' attrition rate (VAR)[21], but not through 'involuntary attrition' caused by redundancies or firings.

---

[19] If the direct/indirect workforce figures did not match the total workforce in the shipyard surveys, the proportions of direct/indirect workers were assumed to be correct and were applied to the total workforce reported in the surveys.

[20] In the model, workers retire according to the workforce age distribution and the average retirement age reported by the shipyard. If the figures of the workforce age distribution did not match the total workforce, the proportions of workers in each age range were assumed to be correct and they were applied to the total workforce reported in the survey. In addition, whenever shipyards did not report an average retirement age, it was assumed to be 65.

Because shipyards reported the age distribution only by ranges (i.e., less than 21, 21–30, 31–40, 41–50, 51–60, more than 60), retirements were smoothed to yearly retirements by dividing the total workforce in an age range by the total number of years in that range. To avoid double counting those who already left the workforce through non-retirement attrition, the number of retired per year was reduced through the cumulative attrition rate.

[21] Two voluntary attrition rates were calculated for each shipyard by averaging its historical voluntary attrition: the first for the management/technical category and the second for the manufacturing category.

2. We added apprentices to the projected supply of shipyard workers, based on each shipyard's planned intake of apprentices minus the estimated number of dropouts from the programme.[22]
3. We added to the projected shipyard labour supply (plus apprentices) other potentially available labour from limited pools of skilled workers, such as temporary and unemployed workers, in the United Kingdom's major shipbuilding regions.[23] We did not,

---

[22] To simplify, we assumed that all dropouts occurred after the first year of the apprentice programme. After their first year, apprentices who had not dropped out were moved into the shipyards' regular workforce, but only in the following subcategories of workers: technical, structure, and outfitting. (We did not consider managerial or direct support hires in the apprentice case.) The number of apprentices who move into each of these subcategories was determined by their share of the total workforce. After apprentices have moved into the regular workforce, they are subjected to the same attrition rules as every other worker (explained in the no-hire scenario).

[23] The available pool of workers in each shipbuilding region of the United Kingdom was estimated using two basic methods. In Scotland and northeast England, the available labour supply in 2004 was determined by subtracting the existing labour forces of the naval shipyards in regions from the total number of workers in the shipbuilding-related trade databases provided by Scottish Enterprise and Swan Hunter. Because of the lack of regional trades databases that pertain to northwestern and southern England, another method was used to estimate the size of the available labour pool in these regions, which was based on the number of active unemployed. In northwest England, it was assumed that the number of available unemployed shipbuilders relative to the total number of active unemployed was about the same for this region as it was for northeast England: 6 percent. In southeast and southwest England, we assumed that the percentage of available unemployed shipbuilders relative to the total number of active unemployed was less than that found in northeast and northwest England: 4.5 and 5 percent, respectively. In regions where specific information on the available workforce's skills was lacking, the pools' distribution across subcategories was assumed to be the same as the distribution of the shipyards in the pools' regions. Also, the age distribution of each pool was determined by the age distributions of the shipyards in the pools' regions, as was the average retirement age of available workers. Each pool's workforce attrition was estimated using the same attrition rules that were applied to shipyards in its region. In the case of regions that contain more than one shipyard, the VAR was calculated by the average of the VARs of the shipyards in that pool. Each year, the naval shipyards add workers from their regional pool, and these numbers are not replaced in the pool. If the number of available workers in any pool becomes less than the sum of the recruitment ceilings for the shipyards in that region, shipyards recruit the remaining workers proportionally to the size of their workforces. Once the pool of unemployed workers is zero for any labour subcategory, shipyard recruitment in that subcategory becomes zero from that year on. Workers recruited by a shipyard from the available labour pools retire according to the same retirement rules that apply to the shipyard's original workforce.

however, include redundant workers in the offshore oil industry in northern Scotland.[24]

Based in part on what we have learned from our research on the US shipbuilding industry, we initially decided to restrict the number of a shipyard's annual recruits (apprentices plus available workers) either to the average of the previous five years or to 8 percent of its workforce in the previous year, whichever was less.[25] However, because of uncertainty regarding the ability of UK shipyards to absorb a large number of workers in a short period, we also decided to perform an excursion of the third case in which all the shipyards were allowed to grow at a rate of 8 percent per year.

Shipyard Labour Supply Model

Figure 3.9 provides a graphical summary of the shipbuilding labour supply modelling process. To reiterate, we started with the pool of labour in the United Kingdom's naval shipyards in 2004, subtracted those that were expected to depart the shipyards, added workers expected to complete their apprentice programmes, and then added other potentially available workers with skills relevant to shipbuilding.

**Figure 3.9
Shipyard Labour Supply Model**

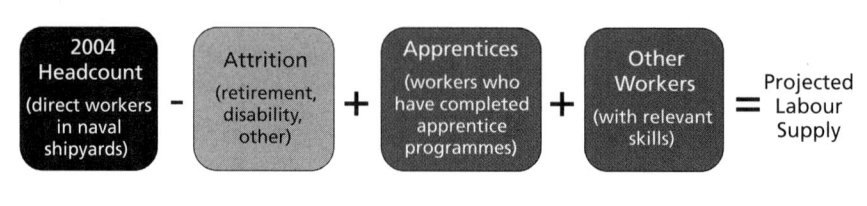

2004 Headcount (direct workers in naval shipyards) − Attrition (retirement, disability, other) + Apprentices (workers who have completed apprentice programmes) + Other Workers (with relevant skills) = Projected Labour Supply

RAND MG294-3.9

---

[24] According to the Highlands and Islands Enterprise skills database, this group consisted of approximately 2,400 workers in February 2003.

[25] In recent years, of the naval shipyards, only Swan Hunter has an average recruitment rate that exceeds 8 percent.

We combined the results of the additions and reductions to the labour force to derive a forecast for a particular year. Then we plotted these outcomes over 17 years to obtain a total picture of the potential naval shipyard labour force in the United Kingdom from 2004 to 2020. Finally, we compared the potential supply picture in the three cases with the projected total demand under the current MOD plan and the level-loading scenario described in the previous chapter (see Figure 3.10).

Results of the Shipyard Supply Analysis

This section shows the results of the three shipyard labour supply cases described above. In the no-recruitment case, we determined that the direct workforce in the naval shipyards could decline by almost

Figure 3.10
A Comparison of the Shipyard Labour Supply and Demand, 2004–2020

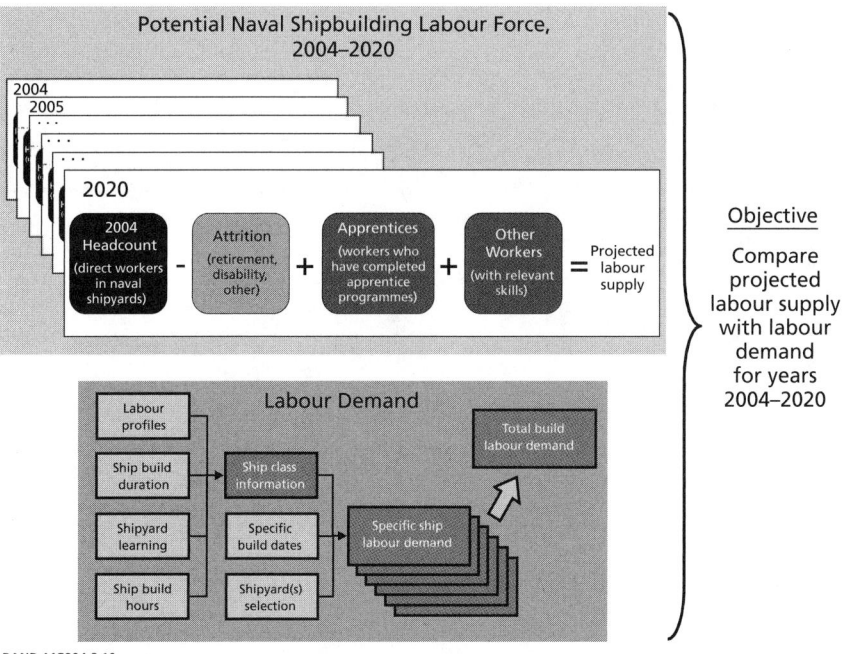

65 percent by 2020 through retirement, incapacity, and voluntary attrition. As Figure 3.11 indicates, the number of workers in the management and technical and manufacturing skill categories would drop from nearly 14,000 workers to around 4,800 in the next 17 years if no steps were taken to replenish the workforce, through hiring either apprentices, temporary workers, older workers from other industries, or going through the ranks of the unemployed.

In Figures 3.12 through 3.14, we compare the potential supply of direct naval shipyard workers in our two recruitment cases with the demand under the current MOD plan and the level-loading scenario. In each figure, the light blue line represents the case in which (after accounting for worker attrition) we added apprentices to the naval shipyards. The grey line represents the projected supply plus apprentices and the pool of available workers with skills relevant to ship-

Figure 3.11
Projected Shipyard Labour Supply by Skill Category Without Additional Recruitment, 2004–2020

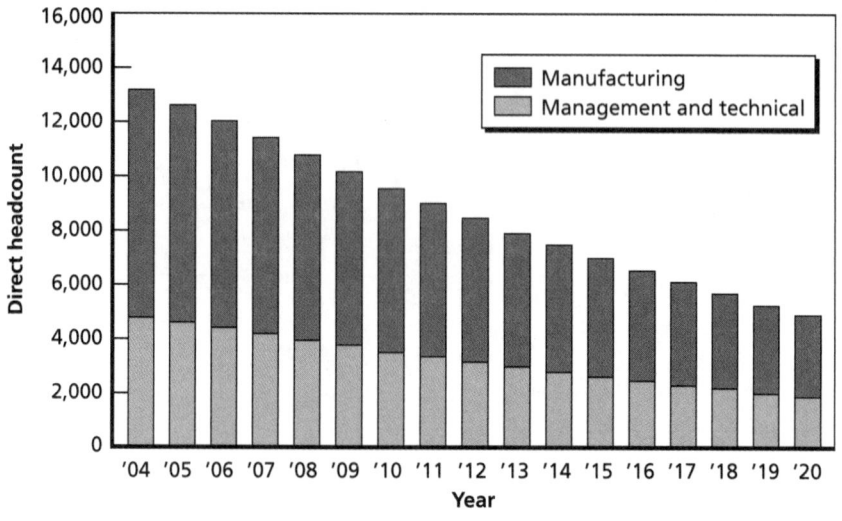

building. The dark blue and black lines represent the industrywide demand for workers under the current MOD plan and our level-loading scenario, respectively.

Figure 3.12 shows the results of the two recruitment cases in the situation in which annual shipyard recruitment is held to the historical average or 8 percent of the previous year, whichever is less. It indicates that the overall supply of management and technical workers would probably be sufficient, even in peak periods, if the naval shipyards were able to recruit from the pool of workers currently outside their industry. However, the naval shipyards may have trouble retaining management and technical workers in periods of decreased demand—for example, from 2004 to 2006 and from 2010 to 2014—particularly under the current MOD plan. Level-loading would reduce the demand during the peak and thus increase the margin of safety in terms of the supply. The drop in near-term demand, however, would still pose a problem for the industry under this scenario.

Figure 3.12
Forecast of the Naval Shipyard Management and Technical Workforce: Minimised Recruitment

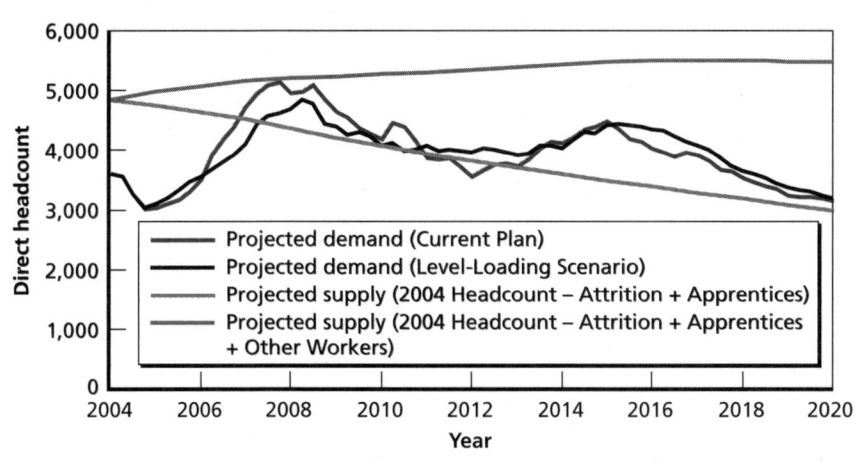

Figure 3.13 shows that the naval shipyards may have trouble meeting the demand for manufacturing workers in the 2007–2011 peak period if annual recruitment is kept to the historical average or 8 percent. This problem will undoubtedly be worse for particular shipyards, especially those in southern England but also in the Glasgow and Tyneside regions. The problem is not so much related to the availability of workers—especially in Scotland—as it is to constraints on the ability of the shipyards to rapidly absorb a large influx of workers in a short amount of time. In BAE's case, the problem could be alleviated by shifting workers from Barrow, where demand is less and supply ample, to the Glasgow yards, which will soon experience a large increase in demand for workers.

It will still be difficult to satisfy the need for manufacturing workers during the peak period even under the levelled demand scenario (see Figure 3.13). However, the supply situation would signifi-

Figure 3.13
Forecast of the Naval Shipyard Manufacturing Workforce: Minimised Recruitment

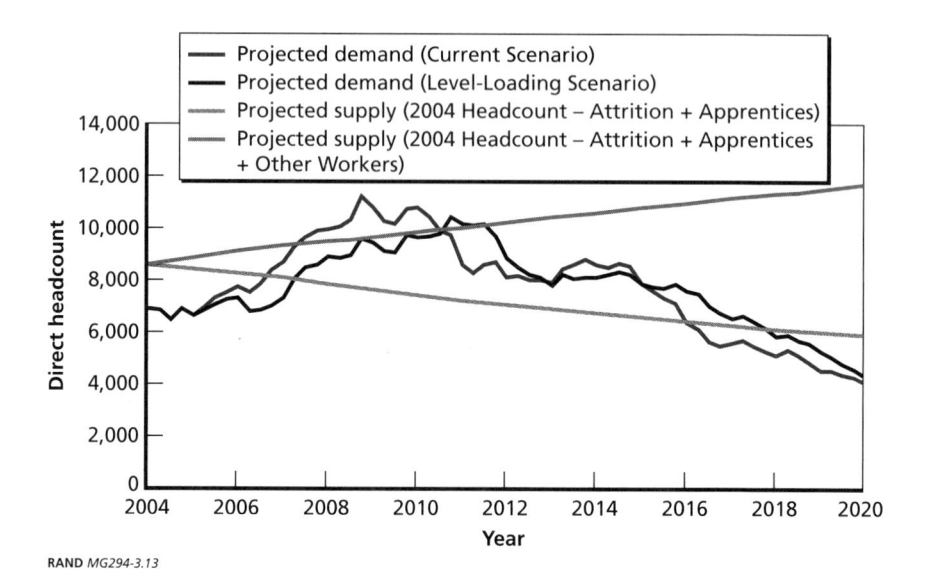

cantly improve over the current demand scenario. In the most optimistic supply case, the current demand scenario would lead to a shortage of more than 1,000 workers in 2009. However, level-loading would reduce the demand-supply gap to fewer than 100 workers in 2011, which could be closed through productivity improvements, increased overtime, and/or outsourcing.

As Figure 3.14 indicates, even the current plan could be achieved (or nearly so) if the naval shipyards were able to expand their workforces at an annual rate of 8 percent. Although a few UK shipyards, such Swan Hunter and DML, have grown at this rate or better in recent years, most major shipyards have traditionally expanded at a slower pace, and it is uncertain whether the entire industry could grow so quickly, especially if the yards had to assimilate a significant

Figure 3.14
Forecast of the Shipbuilding Manufacturing Workforce: 8 Percent Recruitment Rate

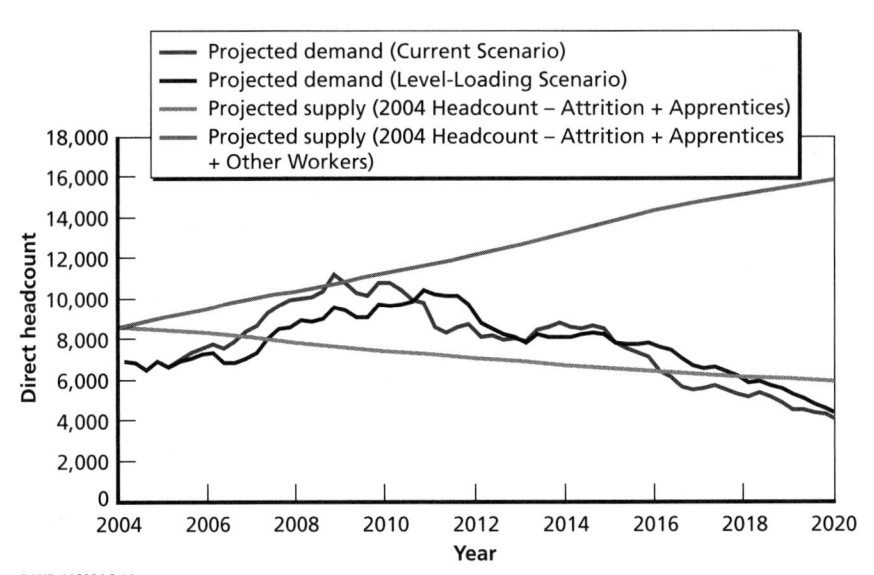

RAND MG294-3.14

number of inexperienced workers into their ranks. That said, our analysis shows that increased work-sharing among the naval shipyards and/or increased interregional mobility of shipyard labour could help to alleviate shortages in specific locations during peak periods, given the variability of potential worker demand and supply in individual shipyards over the next 17 years.[26]

Concluding Observations

Our analysis led us to several concluding observations regarding the United Kingdom's shipbuilding and repair labour supply:

- The number of management/technical and manufacturing workers employed in naval shipbuilding and repair have held relatively steady in recent years, despite the long-term decline in the shipbuilding and repair industry as a whole.
- The naval shipyard labour force is ageing and will severely decline over the next couple of decades without continued, steady recruitment—a trend that UK shipbuilding experts have noted for some time and that our recent shipyard survey confirmed.
- Although there is no immediate labour shortage, the shipyards themselves have told us that they are concerned about the future availability of particular skills (e.g., design, electrical, test and commissioning) as well as the upcoming surge in demand for manufacturing labour, such as steelworkers.
- Despite obstacles to increased recruitment, naval shipyards could potentially tap several pools of labour in related industries, such as offshore construction and commercial shipbuilding, and among the unemployed. They could also continue the trend towards greater outsourcing.

---

[26] Because of business sensitivities, we are unable to provide specific information on the potential demand and supply situations in particular shipyards.

- Acknowledging the limits of our data and the degree of uncertainty regarding the ability of the shipyards to rapidly absorb new workers, we believe the naval shipbuilding and repair industry will have a difficult time meeting the increased demand for warships over the next couple of decades. Levelling the demand can lessen this difficulty, but it will still exist, particularly in certain regions and shipyards, unless work-sharing arrangements are established. At best, meeting the demand will require a high degree of cooperation and work-sharing between the shipyards.

CHAPTER FOUR
Facilities Utilisation at the UK Shipyards

The United Kingdom today is home to a finite number of firms that produce and repair naval ships or submarines. Many other commercial firms in the United Kingdom have facilities that could be used in the future to help accommodate the current and projected programmes. Do these firms have the capacity to produce the future fleet in the time frames specified by the MOD?

The previous two chapters addressed one aspect of this question through an analysis of the labour supply and labour demand. This chapter evaluates how current facilities at UK shipyards may or may not have the capability to produce the future fleet. At any point throughout the build process, production could be slowed or halted as a result of any number of facilities constraints. These potential production limitations could be issues related to throughput or facility availability. Both issues will be explored in this chapter.

Ship Production Facilities and Phases

The production of a ship or submarine involves numerous facilities, including a wide range of shops, cranes, specialised equipment, docks, and piers. Producers employ facilities at different times, in different sequences, and in different ways, depending on the platform being built, yard organisation or layout, build strategy, and many other factors. Figure 4.1 depicts the three production phases we refer to in this analysis, with each phase corresponding to particular facilities it requires. These are not the traditional shipbuilding phases commonly known and used throughout the industry (i.e., design, production, outfitting, test, commissioning, and trials). Nevertheless, we defined these phases because the traditional categorisations did not allow us to adequately map the use of particular facilities. Figure 4.1's black area corresponds to the pre–final assembly phase, the light blue to the final assembly (FA) phase, and the dark blue to the afloat outfitting (AO) phase.

The pre–final assembly phase entails a manufacturing period before final assembly of blocks and modules begins, that is, the period before the ship occupies an assembly location. During this time, facilities such as pipe fabrication shops, unit assembly areas, lay-down areas, and steel fabrication shops are used. Final assembly begins when a producer starts assembling the ship using a facility such as a dry dock, floating dock, slipway, land-level area, or ship assembly hall. Afloat outfitting begins when a ship is launched or floated and ends when the ship is delivered. A ship in the AO phase would require a pier, quay, lock, or a dock.

There is some overlap in use of different facilities throughout each phase. In many cases, certain facilities—cranes, shops, or fabrica-

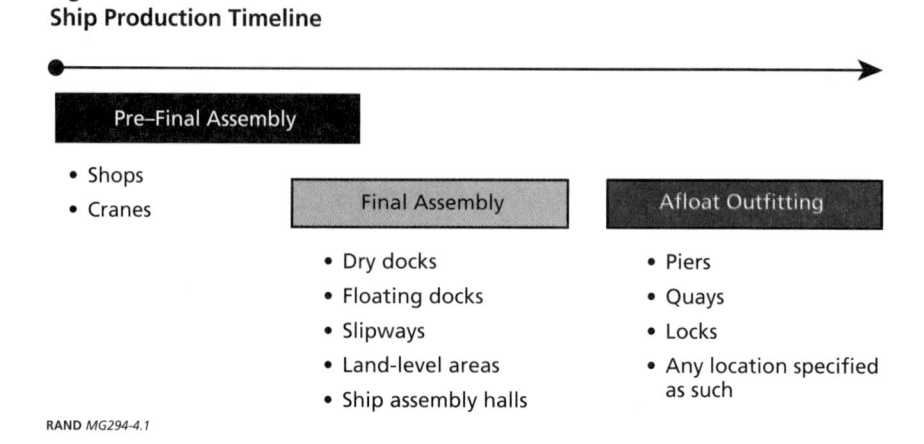

Figure 4.1
Ship Production Timeline

Pre–Final Assembly
- Shops
- Cranes

Final Assembly
- Dry docks
- Floating docks
- Slipways
- Land-level areas
- Ship assembly halls

Afloat Outfitting
- Piers
- Quays
- Locks
- Any location specified as such

tion facilities associated with the pre–final assembly—are used throughout the FA and AO phases. Generally, the FA and AO phases are mutually exclusive, but sometimes an FA facility will be used for outfitting.

How We Studied Facilities and Phases

We focused our evaluation of facility throughput on the FA and AO facilities for two main reasons:

1. There was a lack of consistent measures of throughput for the other types of facilities. For facilities such as shops, it is very difficult to devise objective measures of throughput. The number of pipes that a pipe shop can manufacture, for example, depends on the complexity of each unit, length, diameter, number of bends, etc. Thus, simply stating the number of pipes per day as capacity could be misleading.
2. Consistent measures of throughput would require a prohibitive amount of data from the shipyards. For the example of a crane, the throughput depends on where the crane is located in the yard, what the build strategy is of the programme that will utilise that crane, and how easily the crane can be moved or how easily supplemental cranes can be brought in. Each unit lift must be tracked and assessed. This information would be needed for each crane, lift, and vessel in the yard. Such a data collection was beyond the scope and means of this study and would have placed an undue burden on the shipyards providing information.

As part of the broader survey to the shipyards, we requested information on the number and types of their facilities and on the timings of the programmes they were or are expecting to be engaged in. Each shipyard identified its specific facilities related to ship production. The shipyards also provided the size of the largest ship that each facility could accommodate (length, beam, and draught). The firms included as part of this survey are BAE Systems, BES, DML,

Ferguson, FSL, KBR Caledonia, Swan Hunter, and VT Shipbuilding. In addition, we included facility information based on prior work for the MOD. These facilities included those from A&P Group, some facilities associated with the former Cammel Laird yards, and Harland and Wolff.

Identifying Demand and Assigning Facilities to Phases

To determine whether there were potential capacity limitations of the FA and AO facilities, we first had to assess when programmes would require particular facilities and compare these demands to the number of facilities available. We split each programme into the three phases identified in Figure 4.1. Then, we identified the specific facilities that would accommodate final assembly and afloat outfitting for each programme. Next, we calculated the total demand for each of these facilities on a quarterly basis and compared it with the capacity of those facilities. A detailed example of how this analysis unfolds will be shown later in the chapter.

Only a few programmes in the current MOD acquisition plan that we discussed in Chapters One and Two have been allocated to a particular location (or facilities). In some cases, such as for the Type 45, we know (for the most part) which facilities the programme will use. In other cases, such as for the CVF, we have only notional ideas about the build strategy and the possible allocation of work among the shipyards. In other instances, such as for MARS, we do not have definitive information of the dimensions of ships that will be built or the shipyards that might be involved in construction. Therefore, for the programmes with more defined information, we can undertake a more detailed evaluation.

For programmes with little definitive information, our results are speculative, and we explored multiple, possible locations. In cases in which we knew the yard a programme will go to but not necessarily the facility, we made notional allocations based on size constraints: Large ships go to large facilities, small ships to small facilities. In cases in which the yards or IPTs indicated that specific facilities will be used by a programme, we used those assumptions.

Final Assembly Facilities' Capacity and Considerations

The survey responses and supplemental data indicate that there are approximately 48 FA facilities and 37 AO facilities that are currently or could be used to process future production workload.[1] These FA facilities include those that may have planned upgrades to accommodate future work. However, for this analysis, we used the facilities' current sizes, not their new sizes. Not included in this count are new facilities that do not currently exist but which will be built to accommodate future programmes should the yards be assigned the work.

Figure 4.2 shows the distribution of the maximum length of ship that can be accommodated by a particular FA facility. The bars represent the number of facilities in different ranges of maximum

Figure 4.2
Distribution of Final Assembly Facility Lengths

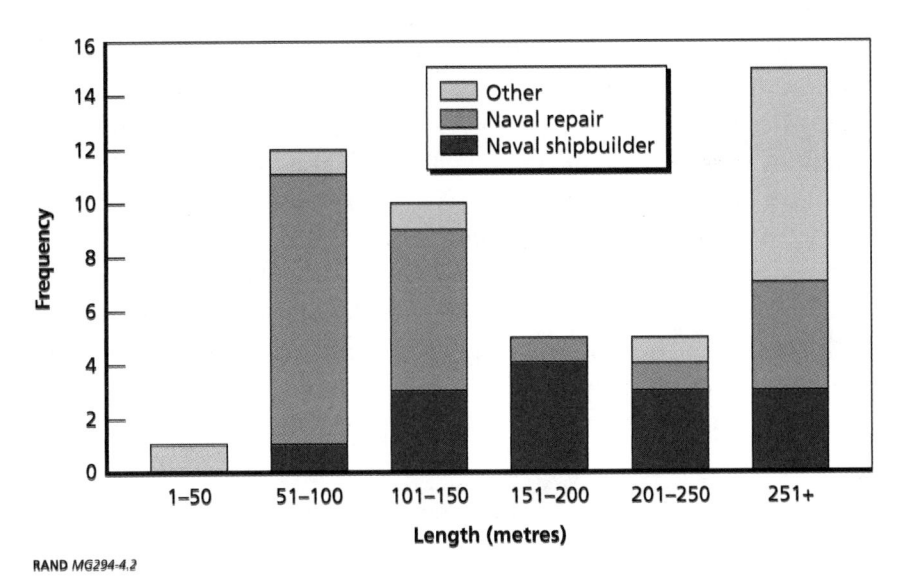

---

[1] It should be stressed that this sample is smaller than the total available in the United Kingdom.

length. The data are further segregated into three categories by the owner of the facility:

- Naval shipbuilder—facilities owned or leased by BAE Systems, Swan Hunter, and VT Shipbuilding
- Naval repair—facilities owned or leased by BES, DML, and FSL
- Other—facilities owned or leased by other firms.

For example, 20 total facilities in the sample can accommodate vessels longer than 200 metres. Of those 20, six are those of the naval shipbuilders, five are of the naval repair yards, and nine are from the rest of the industry.[2]

Figure 4.3 shows the distribution of the largest beam that can be accommodated by the FA facilities. Larger ships, such as the CVF, will require facilities with beams greater than 30 metres and lengths greater than 200 metres. Some RFA ships, such as *Fort Victoria*, have also required very large FA facilities.

Only the very largest of ships require a draught greater than 12 metres. In most cases, such a deep draught represents a ship in a fully loaded condition. Most naval ship construction programmes require facilities with more-modest draughts. As Figure 4.4 shows, 25 facilities can accommodate ships with draughts between 5 and 10 metres, while 14 facilities can handle ships requiring deeper draughts.[3]

Although the distributions of the individual metrics are insightful, the combination of all three metrics determines whether a facility can be used. If the length, beam, or draught is insufficient, the vessel cannot utilise the facility. Table 4.1 depicts a matrix, showing the

---

[2] It should be noted that the data available for the 'other' firms are heavily weighted towards the larger facilities. So no inferences should be drawn concerning the lack of smaller facilities at the other firms.

[3] Of the 48 FA facilities, two are ship assembly halls and six are refit bays. These eight facilities have height restrictions, but no draught measurement. We did not eliminate these facilities but gave them default values of 'not applicable'.

Figure 4.3
Distribution of Final Assembly Facility Beams

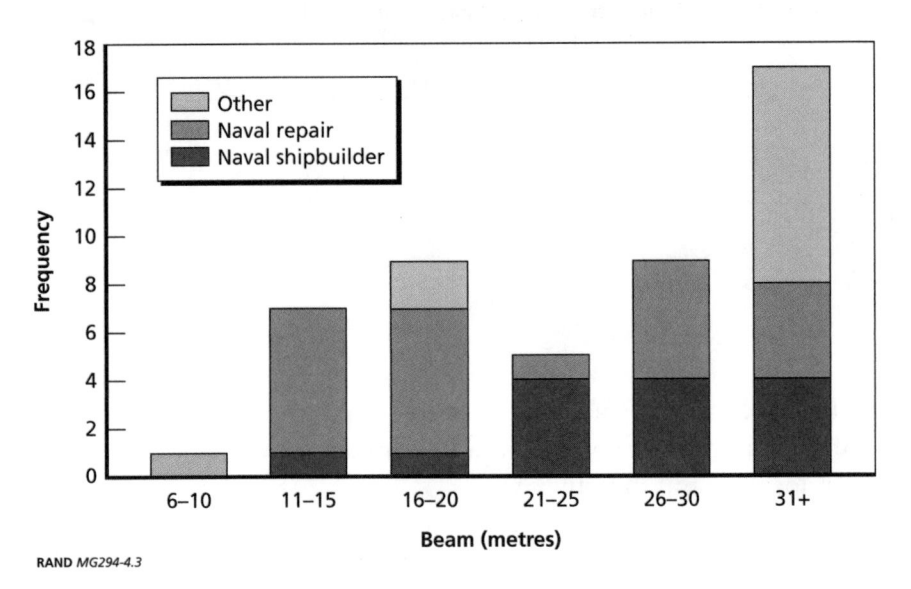

Figure 4.4
Distribution of Final Assembly Facility Draughts

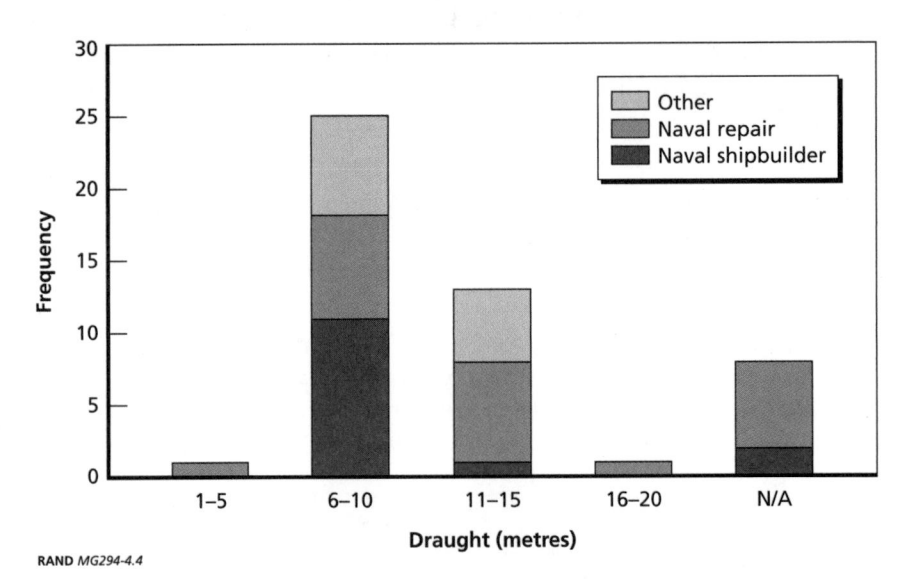

Table 4.1
Number of Final Assembly Facilities at Naval Shipbuilders That Can Accommodate a Ship with a Given Beam and Length

	Vessel Length (m)				
Vessel beam (m)	50	100	150	200	250
10	14	13	10	6	3
15	13	12	10	6	3
20	12	11	9	6	3
25	8	7	6	6	3
30	4	4	4	4	3
35	3	3	3	3	3
40	3	3	3	3	3
45	0	0	0	0	0

number of facilities that can accommodate a ship length of at least the value in the top row, and a ship beam of the value in the left column for the naval shipbuilders.

For example, 14 FA facilities can accommodate a ship with a length of 50 metres and a beam of 10 metres. Conversely, three facilities can accommodate a ship with a length of 200 metres and a beam of 35 metres. It is interesting to note that the median FA facility size at the naval shipbuilders is geared towards a frigate-sized vessel.[4] This fact should not be surprising given the fact that frigates and mine/costal craft comprise the majority of the surface combatant fleet (by number of hulls). Thus, the naval shipbuilding facilities are 'matched' to the size of the existing fleet.

Table 4.2 shows the combined distribution for the naval ship repairers and the other firms. These two groups have several FA facilities that can accommodate larger ships.

Afloat Outfitting Facilities Capacity Considerations

AO facilities follow a slightly different pattern. As shown in Figure 4.5, the distribution of AO facility lengths indicates that nearly half

---

[4] Sizes of these vessels can be found in Appendix B.

Table 4.2
Number of Final Assembly Facilities at Naval Repair and Other Firms That Can Accommodate a Ship with a Given Beam and Length

	Vessel Length (m)				
	50	100	150	200	250
Vessel beam (m) 10	33	22	15	14	12
15	33	22	15	14	12
20	19	18	15	14	12
25	18	17	15	14	12
30	13	13	13	13	11
35	9	9	9	9	9
40	8	8	8	8	8
45	4	4	4	4	4

Figure 4.5
Distribution of Afloat Outfitting Facility Lengths

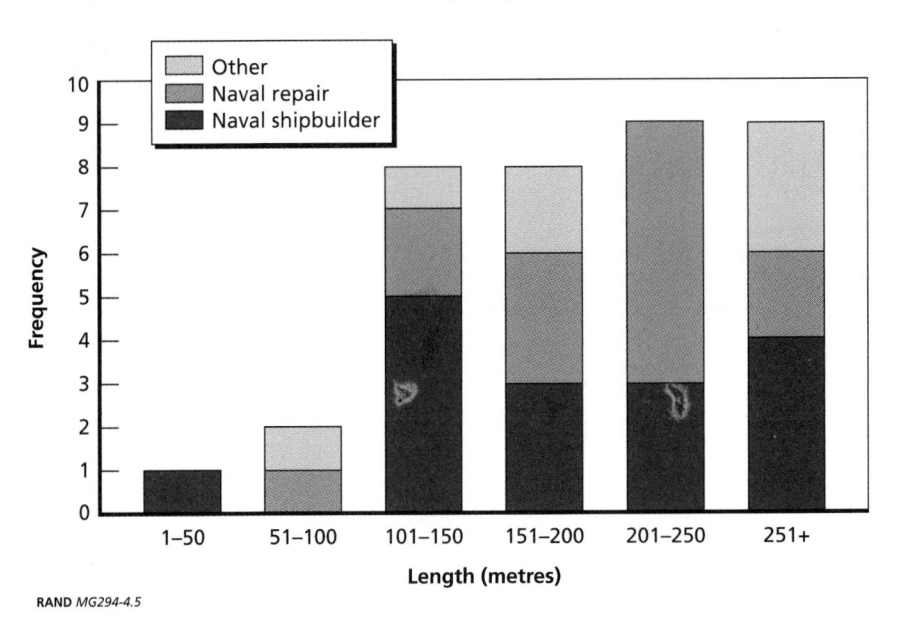

RAND MG294-4.5

have a length greater than 200 metres. Unlike most FA facilities, multiple vessels can easily use a single AO facility at one time. Thus, the longer the length, the more flexible the facility is for afloat outfitting.

98 The United Kingdom's Naval Shipbuilding Industrial Base: The Next 15 Years

Over half the AO facilities have a beam greater than 30 metres (see Figure 4.6). This distribution results from the fact that many outfitting quays lie directly on a river, channel, or harbour, where width is not an issue. Thus, the beam of the ship for AO facilities is not a significant constraint at most locations.

One consequence of having AO facilities at a tidal location is that the maximum draft can be limiting. As Figure 4.7 shows, the mode of the AO facility draught is between six and ten metres. Thirteen of the facilities have a draught capacity of greater than 10 metres, while 19 facilities support draughts of between 6 and 10 metres.

As with FA facilities, the length and beam of AO facilities are highly correlated. Table 4.3 portrays a similar matrix as Table 4.1, showing data for AO facilities at the naval shipbuilders. Table 4.4 displays the same data for the naval repair and other firms. Note that there are only a few AO facilities that have the capacity to accommodate large ships with beams greater than 30 metres and lengths greater than 200 metres. Again, some care must be taken because of draught restrictions.

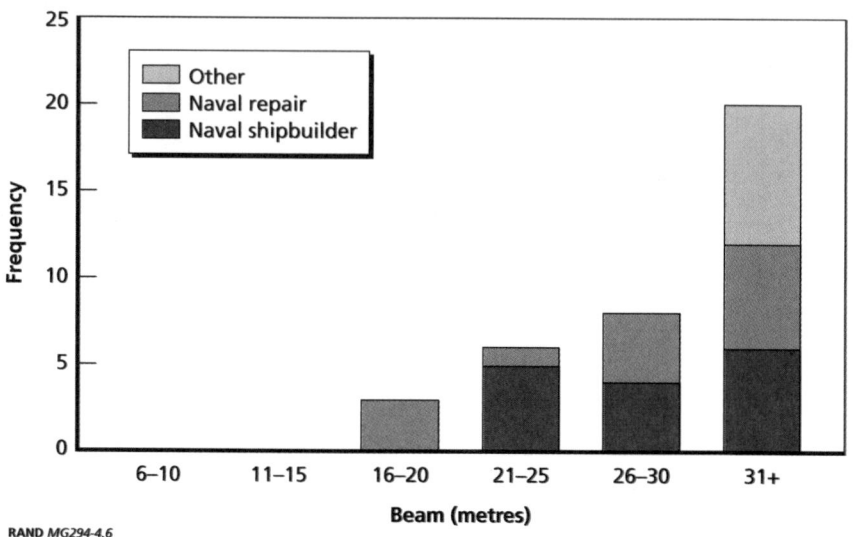

Figure 4.6
Distribution of Afloat Outfitting Facility Beams

Figure 4.7
Distribution of Afloat Outfitting Facility Draughts

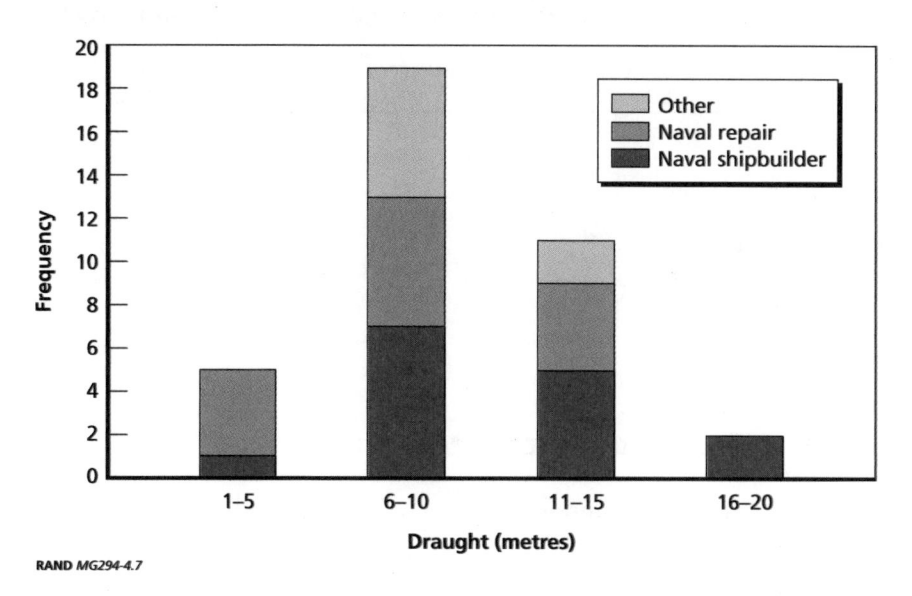

RAND MG294-4.7

Table 4.3
Number of Afloat Outfitting Facilities at Naval Shipbuilders That Can Accommodate a Vessel with Certain Length and Beam Characteristics

	Vessel Length (m)				
Vessel beam (m)	50	100	150	200	250
15	15	15	10	8	4
20	15	15	10	8	4
25	10	10	9	7	3
30	6	6	5	4	1
35	3	3	2	1	0
40	3	3	2	1	0
45	3	3	2	1	0

Table 4.4
Number of Afloat Outfitting Facilities at Naval Repair and Other Firms That Can Accommodate a Vessel with Certain Length and Beam Characteristics

		Vessel Length (m)				
		50	100	150	200	250
Vessel beam (m)	15	21	19	16	11	5
	20	21	19	16	11	5
	25	18	16	15	11	5
	30	15	14	13	9	5
	35	13	12	11	8	4
	40	11	10	9	7	3
	45	11	10	9	7	3

Capacity Implications for Future Programmes

Much of the data we used for this analysis are sensitive and are not for publication. We performed our analysis on a yard-by-yard basis and results and conclusions were specific to facilities and shipyards. All the workload being placed on a particular yard was compared with the capacity of the facilities within that yard.

Our analysis reveals two main causes of potential facilities problems in the naval shipyards: One problem was caused by a lack of FA or AO facilities of a particular size; the other was caused when the demand for a particular sized facility exceeded its capacity. In some cases, there was simultaneous demand from several ships for a single facility that could only accommodate one ship at a time. The remainder of this section will discuss the implications of facilities for each of the major future programmes that will place a demand on these 48 FA and 37 AO facilities.

We will show the detail of how this analysis was performed only for the first programme: the Type 45.[5] For the remaining cases, we will present only the conclusions and highlights of the analysis.

---

[5] In this case, the requirements for FA and AO facilities for ships one through six were laid out. Then, we observed capacity requirements for FA and AO facilities on a quarterly basis. (The capacity requirement refers to the total capacity required to accommodate demand in the yard. If three ships each require an FA facility, then the capacity requirement is three.)

Type 45

As noted, there is currently a contract for six Type 45 ships with BAE Systems and VT producing the ships. We show the planned distribution of work between the shipbuilders in Figure 4.8.[6] VT Shipbuilding will build two blocks of the ship (E and F) as well as the superstructure. The BAE Systems Clyde yards will be responsible for the remaining block structures.

The VT Type 45 blocks require an assembly location for manufacture. VT has indicated that the blocks will be constructed in its new ship assembly hall, which has two build lines, each with different dimensions and therefore different capacities. The first build line has a capacity for up to two Type 45 bow sections at the same time (in

Figure 4.8
Type 45 Work Allocation

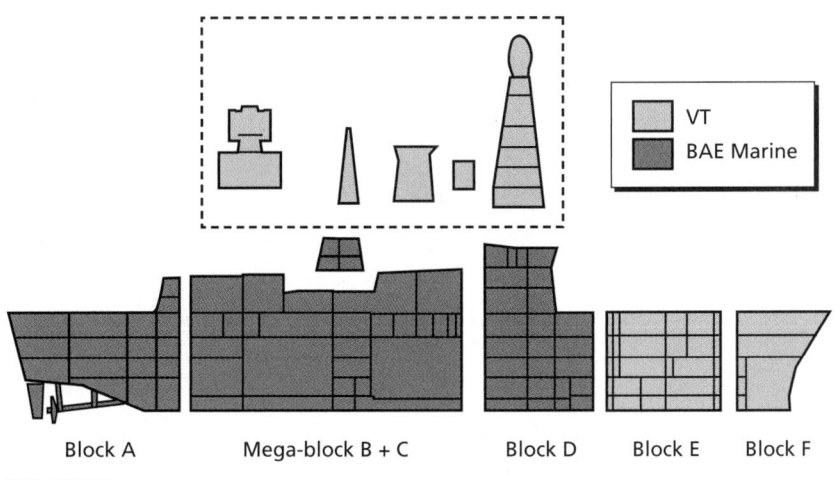

This allowed us to compare the capacity requirement with the capacity of either the facilities specified as potential Type 45 facilities or facilities that were of the size to accommodate such a programme.

[6] Birkler et al. (2002).

different stages of build); the other line has a capacity for approximately one ship of OPV(H)/Corvette size.[7]

The first-of-class Type 45 will be assembled and launched at Scotstoun, up on the Clyde. The remaining ships will be assembled and launched at Govan and Scotstoun on the Clyde.[8] FA and AO phases of the build at BAE will require facilities that can accommodate a ship with a length of 152.4 metres, a beam of 21.2 metres, and a draught of 5.0 metres.[9] Three facilities on the Clyde, at Scotstoun and Govan, can accommodate a ship of this size. Two of these facilities are docks, and the other is a slipway. We do not have detailed plans of how these three facilities will be utilised for Type 45 production.

To determine whether there would be any complications related to these facilities, we laid out Type 45 facilities requirements over the time periods in which an FA and AO facility will be required at VT and on the Clyde. For example, in the fourth quarter of 2005, there are blocks for three hulls being built by VT. We compared these requirements with the facilities that are available to accommodate these requirements. The prospective build location at VT—the ship assembly hall—has the capacity in build line one for up to two Type 45 bow sections, and the other build line has the capacity for a whole ship of the size of a Future Offshore Patrol Craft, or Corvette. We show the total number of FA facilities capacity required, by quarter, at VT in Figure 4.9.

The maximum number of facilities required in any given quarter is three, but generally, the demand is for one or two facilities. Two things should be noted in this data display. First, Figure 4.9 shows only the facility demand that is coming from the Type 45 programme. Any other work in the yard that would place a demand on

---

[7] As indicated in survey response.

[8] These dates as well as much more information about the Type 45 programme can be obtained at the following Type 45 IPT Web site: www.baesystems.com/type45/index.htm (last accessed November 2004).

[9] Dimensions taken from Royal Navy (2003).

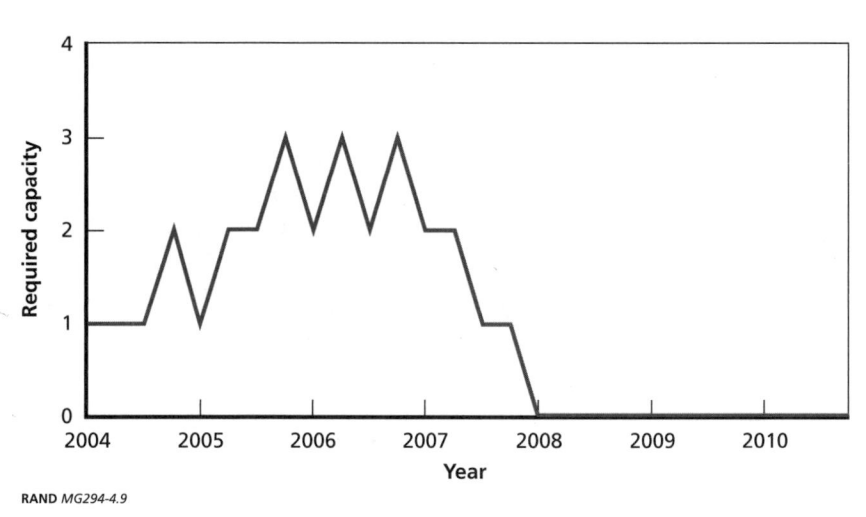

Figure 4.9
Final Assembly Facilities Requirements for Type 45 Programme: VT Shipbuilding

the ship assembly hall at the same time as the Type 45 block builds is not shown. Second, changing the resolution of the time measurement from quarters to months may alleviate some of the single quarter spikes in required capacity. That is, the requirement for a facility could start or end midway through a quarter and not actually overlap with another ship.

If both assembly lines in the ship assembly hall are used to construct the Type 45, these brief periods of demand for a capacity of three can be easily accommodated. However, if only the assembly line (which can accommodate up to two Type 45 bow sections) is used, then the yard may not be able to accommodate the periods in which there is a brief capacity requirement of three.

The FA facilities requirement at BAE is similar to the requirement at VT. As shown in Figure 4.10, there are three quarters in which there is a requirement for three FA locations each with a capacity for one ship, or one FA location that has the capacity for three ships. In most quarters, the capacity requirement is for one or two FA locations.

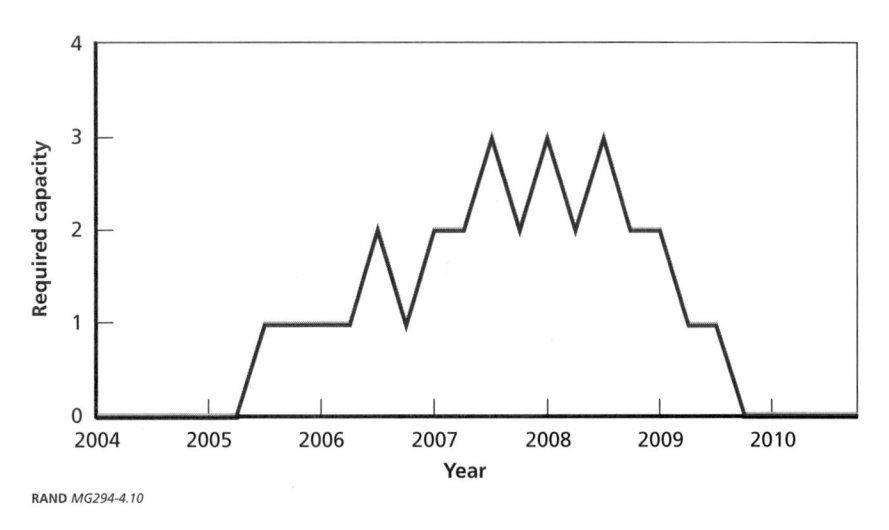

Figure 4.10
Final Assembly Facilities Requirements for Type 45 Programme: BAE Systems

RAND MG294-4.10

There are three FA locations on the Clyde, each large enough to handle one Type 45. It is unclear how these facilities will be utilised or if there are other factors such as craneage or logistics complications that would prevent any of these facilities from being used. However, each of the three facilities is large enough to accommodate a Type 45 vessel. If all three facilities are used to build these vessels, then the FA facilities will not be a limiting factor. The total number of available facilities that could accommodate the Type 45 builds in any quarter is equal to or greater than the maximum demand in all quarters. If fewer than all three facilities can be used, the facilities would limit throughput of Type 45 to less than that planned.

As mentioned above, it is possible to use FA facilities for afloat outfitting as long as the ship can easily come in and out of the facility. For example, a ship could be launched from a dry dock, floated out, tested, and returned to the dock for outfitting. However, this is not feasible to do on a slipway.

The AO facilities as defined in this text are not required at VT for the Type 45 programme because the blocks will be floated out on barges and shipped to the Clyde yards. However, there is a requirement for these facilities at the Clyde yards. The AO period in Figure 4.11 includes some sea trials in which the facility will not be utilised. However, the facility needs to be available for the duration of the period. The figure shows the number of facilities (each with a capacity for one ship) that will be required for afloat outfitting, in each quarter.

In several quarters, the maximum requirement is for two AO facilities, which, again, is the facilities demand from the Type 45 programme only. Any other demand in the yard for the AO facilities would increase the total demand by quarter. There is likely to be two AO facilities available, each with the capacity for one Type 45.[10] One of these facilities may be unavailable because of other work to be

Figure 4.11
Afloat Outfitting Facilities Requirements for Type 45 Programme: BAE Systems

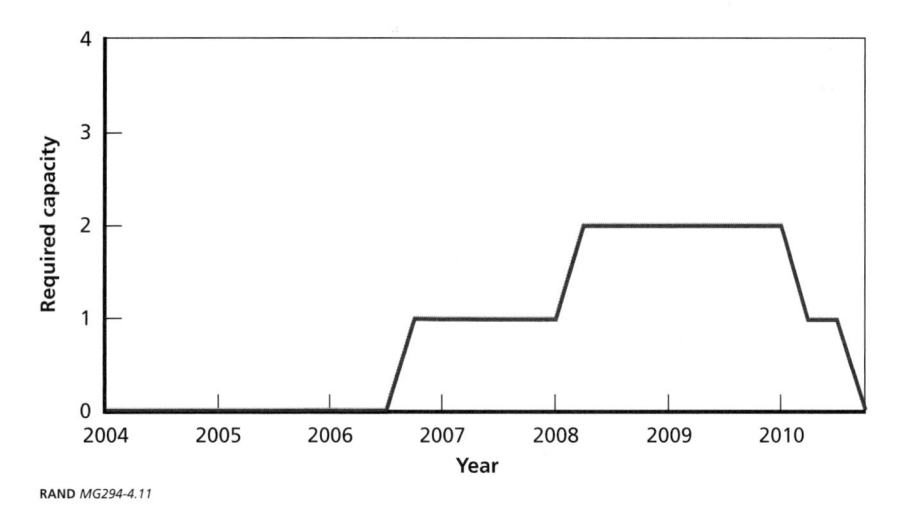

---

[10] This assumes that other AO facilities at the shipyards are converted for CVF production.

done in the yard, leaving one facility to accommodate the AO demand. If docks are used as AO facilities, in addition to their FA role, the total capacity for outfitting of Type 45 would exceed the maximum demand in each quarter. However, a detailed scheduling analysis would be required to determine whether such a problem actually exists and to suggest mitigation solutions.

CVF

There are currently plans to build two new aircraft carriers to replace the retiring *Invincible*-class carriers. Although the design, size, and build strategy of these new carriers have not yet been specified, the MOD has stated that

> initial indications suggest that the carriers could be amongst the largest warships ever built for the [Royal Navy]. Studies to date indicate that potentially a combination of four UK yards (BAE Systems at Govan, Babcock BES at Rosyth, Swan Hunter in the North East and VT Shipbuilding in Portsmouth) offers the best way forward for the build of the carriers, although the use of other UK yards has not yet been ruled out.[11]

If CVF work is split between the yards, then some form of a block build strategy would likely be employed. The proportion of work that would go to any of the possible locations is unknown, but the size of the blocks could be significant given the potential size of the ship. Overlapping block builds may pose a potential problem for yards. Our survey indicates that many yards would have to build new facilities or upgrade existing facilities to accommodate the manufacture of blocks. It is unclear whether all the yards have the capability to make upgrades such that multiple simultaneous blocks can be accommodated.

In addition to overlapping block builds, overlaps of integration and block builds could pose capacity problems. If the same facility

---

[11] This information comes from the MOD's Web site ('Project Fact Files: Future Aircraft Carrier [CVF]', updated 28 September 2004, www.mod.uk/dpa/projects/cvf.htm, last accessed November 2004).

used to assemble the CVF is also required to assemble a block, the current schedule may have to be modified. However, creative solutions could be employed to resolve these types of potential conflicts, such as storing blocks on-site until they are required for integration or utilising two adjacent docks and floating the block to the integration location when completed.

The CVF could be one of the largest warships ever built in the United Kingdom. Few FA and AO facilities could currently accommodate such a ship, and these facilities all would require some kind of investment, upgrade, and/or reactivation.

MARS

The MARS programme is pre–Main Gate. As such, the exact quantity and sizes of the ships have not yet been formally stated. To evaluate future demand on FA and AO facilities, we will use as a proxy the numbers and dimensions of the RFA ships that MARS will replace.

According to the Royal Navy, the RFA ships up for replacement 'include ageing Rover class small fleet tankers, some of the larger support tankers and the stores ships RFA *Fort Austin* and RFA *Fort Rosalie*.'[12]

Table 4.5 displays the dimensions of these ships the MARS programme will replace. As an upper bound on the size of any potential ship is the 'Panamax' capable size, that is, the maximum sized ship that can fit through the Panama Canal. The maximum length of a ship that has ever passed through the Panama Canal was 299 metres. The maximum beam of a ship ever to pass through the canal was 32.6 metres.[13]

The possible replacement ships have a wide range of sizes. The specific requirements for length, beam, and draught as well as other UK programme requirements will determine the possible FA and AO facilities that can accommodate the programme. In general, ships that

---

[12] Royal Navy (2003, p. 14).
[13] Lloyd's Register (2004).

Table 4.5
Ships Replaced by MARS

	Length (metres)	Beam (metres)	Draught (metres)
Rover class	140.5	19	7.32
Larger support tankers (*Brambleleaf, Bayleaf, Orangeleaf*)	170.7	26	11
Fort class (*Fort Austin, Fort Rosalie*)	185.1	24	9

have a length greater than 200 metres, a beam greater than 30 metres, and a draught greater than 7 metres will have several choices for final assembly. However, most of these facilities are at a location not currently engaged in shipbuilding. It must also be kept in mind that the MARS ships will have significant cargo capacity; thus, the draughts in a light-ship condition will be less than that in a fully loaded condition (as shown in Table 4.5). In summary, the smaller the replacement for the retiring classes of RFA ships, the more FA facilities there are to accommodate production. AO facilities might be limiting if deep draughts are necessary.

In terms of the number of facilities required for the MARS programme, this number will depend on the number of ships produced, the period needed in an FA location (related to the build duration), and desired delivery schedule. For example, a hypothetical case of 10 ships, taking one to one-and-a-half years to assemble, with a delivery period of one ship per year would equate to roughly one to two ships in assembly, on average, at any given time. Thus, it is likely that at least two assembly points will be needed for MARS.

Astute

BAE Systems serves as the prime contractor for the *Astute* class and will build the vessels at its Barrow facility, the home of BAE Systems Submarines. The facility that will be used for the FA phase of the build is the Devonshire Dock Hall (DDH). The DDH has two build lines, with a total maximum capacity of four submarines in progress. The DDH's optimal capacity is approximately three submarines in different stages of the build process.[14] Current and future build plans do not place a requirement of more than this capacity on the DDH in any quarter. There is ample AO facility capacity at Barrow for the programme.

LSD(A)

Four LSD(A) ships have been procured. Swan Hunter is building two at its Wallsend location, and BAE Systems is constructing the other two on the Clyde. The LSD(A) requires FA and AO facilities that can accommodate a length of 176 metres, a beam of 26.4 metres, and a draught of 5.8 metres.

The Swan Hunter Wallsend location has one FA facility and ample AO facility capacity for the LSD(A).

BAE Systems has two FA facilities that can accommodate the LSD(A) programme on the Clyde. The LSD(A) is scheduled to be built before the Type 45. The first ship being built at BAE, *Mounts Bay*, is scheduled to be launched in early April 2004, after which the final assembly of the *Cardigan Bay* will begin.[15]

If the last LSD(A) is launched during the same period in which the CVF and Type 45 are being constructed (and that the Type 45 is built according to the stated timetable) there could be a capacity problem with AO facilities at BAE, unless available docks are also used for outfitting purposes. The utilisation of these FA and AO facility resources will have to be carefully planned and scheduled to balance the work in the yard.

Future Surface Combatant

The FSC, which will replace the retiring Type 22 and Type 23 ships, is still in the concept phase, so its design and construction start dates serve as rough estimates. No work has been assigned to any particular yard; however, it is possible that the potential number of these ships

---

[14] As reported by BAE in its survey response.
[15] 'Mounts Bay Awaits Easter Launch Date' (2004).

planned for production could substantially affect FA and AO facility availability.

In our analysis, we assumed that the FSC would be about the same size as the Type 23, with a length of approximately 133 metres, a beam of approximately 16.1 metres, and a draught of approximately 5 metres. A ship with these dimensions would have many facilities choices. About half of the FA facilities and most of the AO facilities could be used. Whether there is a facilities capacity problem will depend on the total demand for the facilities in the United Kingdom at the time of production and the number of producers. However, based on the current plan, the demand for production facilities will generally be low when the FSC is in production and facility availability should not be an issue.

Joint Casualty Treatment Ship

The JCTS will replace the RFA *Argus*. Because the JCTS does not yet have a specific design, we used RFA *Argus* dimensions to assess capacity of FA and AO facilities. The length, beam, and draught, are 175.1, 30.4, and 8.1 metres, respectively. About a third of the FA facilities could accommodate a ship with these size characteristics.[16] However, most of these facilities are not at one of the naval shipbuilders. Eight of the AO facilities can accommodate a ship that size.

Summary

Whether there will be a facility problem depends on the size characteristics of the ship, which yard(s) the programme goes to, and the other demands that will be placed on the same facilities in that yard(s) at the same time. If two ships need to simultaneously use a specific resource, there will be a capacity problem. If the two ships can use the facility sequentially, there will not be a capacity problem.

---

[16] The number of possible FA facilities for JCTS could be greater as the light-ship draught of the vessel will likely be less than 8.1 metres.

Figure 4.12 shows a global view of the facilities demand in each quarter across for the MOD ship programmes.[17] The maximum number of facilities required in any given quarter is less than the total number of facilities available for both FA and AO facilities.

Despite the fact that there appears to be an excess of facilities, a capacity problem could still occur. Again, the size of the vessel and build strategy restricts the location for afloat outfitting and final assembly. Much of the future programmatic demand is for larger ships, which can be accommodated at only a few FA or AO locations. However, a high total demand in each quarter does not necessarily imply a potential capacity problem. There could be a demand for 16 facilities in a single quarter, but if the demand were based on small- to medium-sized vessels, there would likely be no capacity problem.

Figure 4.12
Final Assembly and Afloat Outfitting Facilities Requirements for MOD Ships, 2004–2020

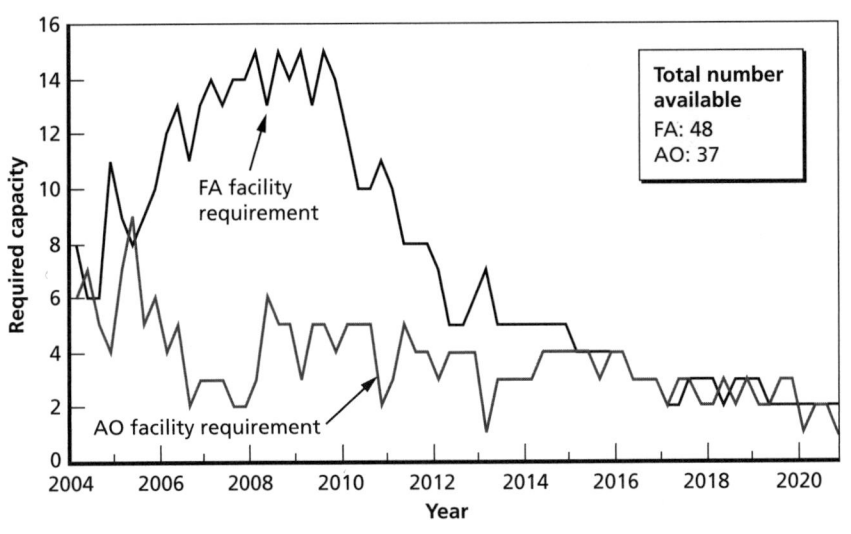

RAND MG294-4.12

---

[17] Note that the demand does not include all refit, refuelling, and repair work. These data were considered sensitive and thus not available to this study.

Conversely, the demand for just two facilities in one quarter could be problematic if the ships were of such a size that only one facility could accommodate the ships.

For the current and planned MOD programme, the total number of FA and AO facilities appears, on the surface, adequate to handle the demand. However, careful scheduling of these facilities will be critical to prevent bottlenecks that may cause schedule delays. More importantly, there are far fewer facilities options for larger vessels. Most of these facilities that can handle large ships are not currently being used for shipbuilding. Some are inactive, while others are used for either naval or commercial repair. This fact will be a particularly important consideration for programmes such as CVF, JCTS, and MARS, which will require that some facilities upgrades and improvements be made as well as the potential reopening of some inactive facilities. Further, it is conceivable that some of the commercial facilities may not be available for use.

CHAPTER FIVE
The UK Shipbuilding Supplier Industrial Base

Shipbuilders do not perform all the work necessary in ship construction themselves. Purchased materials and equipment make up a large portion of a ship, typically more than 50 percent. The shipyards buy equipment from specialised suppliers that may have an expertise in particular areas—for example, communications systems or saltwater desalination systems—which can do the work more efficiently. This ratio is typical of major military weapon systems such as airplanes.

Therefore, as part of this study, we conducted research on the strength of the UK shipbuilding supplier base. We wanted to know if there were any major concerns affecting the various sectors supporting the shipbuilders. A reasonable question is whether the shipyard suppliers will be able to support MOD's future needs. We assessed the shipyards' perspective on the strength of their suppliers and also obtained perspective from the suppliers themselves. Our bottom-line question was whether and to what extent the MOD needs to be concerned about the supplier industrial base.

Research Approach

To get at the question of the strength of the supplier base, we took a two-tiered approach, sending out a main survey to the shipyards and a secondary survey to the suppliers. In the main survey that we sent to the shipyards, we asked for information on their major first-tier suppliers and their critical second-tier suppliers. In this survey, we asked for the names of the suppliers, information on what they provided, and the amount of money spent on them. We also asked the shipyards three subjective questions formulated to get at three different aspects of supplier strength. We asked them (1) whether the suppliers had the capacity to take on more work (rated low, medium, or high), (2) for an assessment of their long-term stability and viability (rated low, medium, or high), and (3) whether there were alternative suppliers for what they provided or whether the shipyard was dependent on that particular company (reverse coded so that riskier situations were rated 'high').

The survey technique generated a list of 349 suppliers. There were many duplicate company listings, which we consolidated if the sites had the same address. In a number of cases, the shipyards purchased goods and services from other shipyards that were also in the original sample, so we did not survey these firms a second time. And there were a few cases in which no contact information was provided.

This consolidation resulted in a second sample of 230 suppliers,[1] to which we sent a short survey assessing other measures of supplier strength (by mail in December 2003). In the packet, we included an introductory description of the study as well as a letter from the DPA asking suppliers to support the research effort. In cases in which we did not get a response within a few weeks, we followed up in January and February 2004 with email and telephone calls if we had the appropriate contact information. As of April 1, 2004, we received answers from 48 suppliers, a response rate of 21 percent, which was in line with the 20 percent reported by the DTI in its 2001 competitive analysis of the UK Marine Equipment Sector.[2]

In the second survey, we asked the suppliers for information on whether they were independent or subsidiaries of other firms, their

---

[1] Realistically, the actual sample size is some number below 230. To increase the potential response rate, we sent surveys to suppliers that were branches of the same company but had different addresses. Some of these indicated that their headquarters would provide consolidated responses for all their locations.

[2] Department of Trade and Industry Engineering Industries Directorate (2001).

dependence on MOD naval work and about the rest of their customer base, and some labour force and recruiting issues.

It should be noted that this sampling frame excludes most government-furnished equipment. The MOD should have more insight into these suppliers because it does business directly with them.

Characterising the Supplier Base—The Shipyard Perspective

We first looked at characteristics of the suppliers that the shipbuilders identified as key first- and second-tier participants in the supply chain.

What They Supply

We asked the shipyards to identify their major first-tier suppliers by industrial sector and their most critical second-tier suppliers. The categories we asked for are as follows:

- Hull structure—assembled main hull body with all structure subdivision
- Propulsion plant—major components installed for propulsion and related systems
- Electric plant—power-generating and distribution systems for ship service
- Command and surveillance—equipment and systems installed to receive, transmit, and distribute information on- and off-ship
- Auxiliary systems—systems required for ship control, safety, provisioning, and habitability
- Outfit and furnishings—outfit equipment and furnishings required for habitability and operability not included in other elements
- Armament—complex of armament and related ammunition and cargo munitions handling, stowage, and support facilities

- Integration and engineering—engineering effort and related material associated with the design, development, production planning and control, and rework of the ship
- Ship assembly and support services—efforts and material associated with the construction that cannot be identified with other elements
- Critical second tier—suppliers that are not first-tier but supply a critical component that is not easily replaced or substituted.

After some research on the businesses of the second-tier companies, we were able to assign them to industrial sectors, which we display in Figure 5.1.

Figure 5.1
Industrial Sectors of the Identified Suppliers

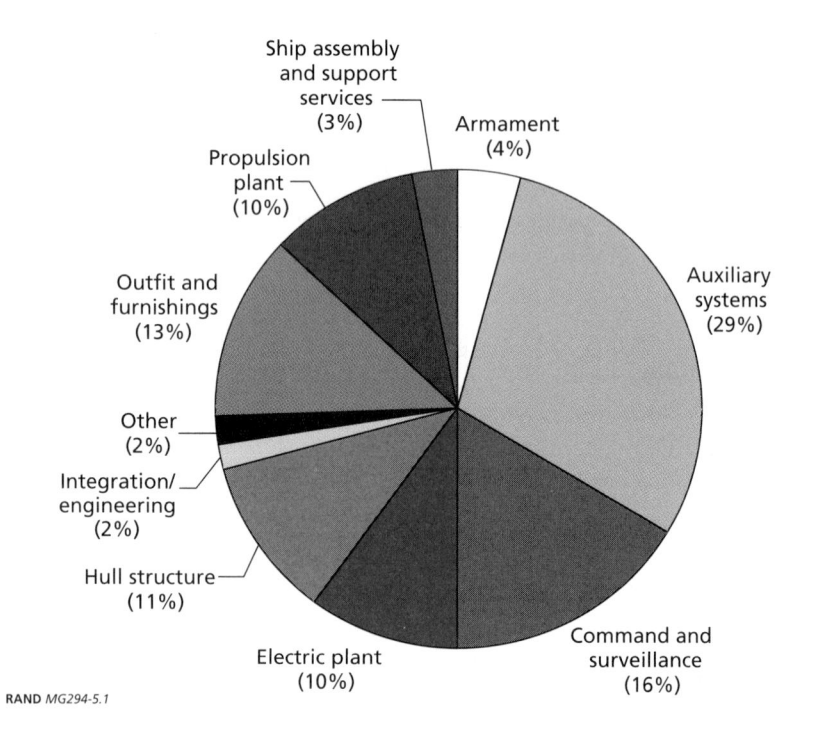

- Ship assembly and support services (3%)
- Armament (4%)
- Propulsion plant (10%)
- Outfit and furnishings (13%)
- Other (2%)
- Integration/engineering (2%)
- Hull structure (11%)
- Electric plant (10%)
- Command and surveillance (16%)
- Auxiliary systems (29%)

RAND MG294-5.1

The figure demonstrates that the identified suppliers support the shipyards in a wide variety of areas, as is to be expected.

The financial data from the shipyards on how much they pay to suppliers was not consistent enough to present here, but it ranged from a few tens of thousands of pounds to many millions.

Where They Are

The shipyards in this study predominantly used suppliers that are based in the United Kingdom. As Figure 5.2 shows, 81 percent of suppliers identified as key first- or second-tier participants were located in England, Wales, Scotland, and Northern Ireland. Following the United Kingdom was the Netherlands, with just fewer than 6 percent of the suppliers. Germany provided slightly fewer than 5 percent, the United States 3 percent, France about 2 percent, and Italy slightly more than 1 percent.

Figure 5.2
Locations of Suppliers, by Country

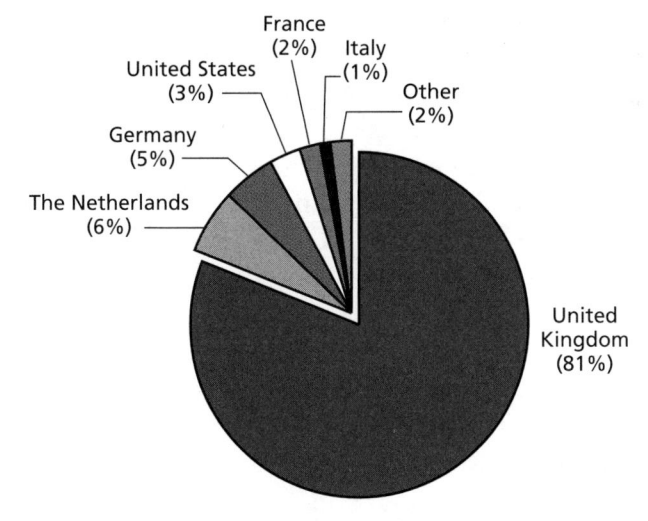

- France (2%)
- Italy (1%)
- United States (3%)
- Other (2%)
- Germany (5%)
- The Netherlands (6%)
- United Kingdom (81%)

RAND *MG294-5.2*

Three Measures of Supplier Strength

To determine the shipyards' assessments of their suppliers' ability to support MOD's future shipbuilding needs, we asked them to report on three different measures of their key suppliers' strength. The first question was the ability of the suppliers to take on more work, the goal of which was to measure whether there was some amount of excess capacity or ability to be flexible that the shipyards could count on, if necessary. The second question was whether the suppliers were likely to be around in the long term, since we wanted to get a sense of whether the suppliers would be available for future MOD programmes. The third question referred to concerns about dependence on particular suppliers and the existence of competition for whatever goods or services they provided. We will examine each measure in turn, breaking out the information by industrial sector in which the suppliers operate.

The first measure is the ability of the supplier to take on more work—that is, its ability to increase production. We display the results of our inquiry in Figure 5.3. Understanding capacity is useful if the customer either increases the total volume of work or re-programs the work so that the peaks increase in magnitude. In the research at hand, extra capacity should give the MOD some comfort that the supplier base would be able to meet its needs if it changes the acquisition schedules for new ship programmes.

Looking at all the suppliers for which the shipyards provided information, we can see in Figure 5.3 that for only a very small percentage of suppliers does capacity appear to be an issue. The one industry for which it appears to be the biggest problem (integration and engineering) has only three suppliers identified by the shipyards, of which one is at risk. (There are more suppliers in the sector as a whole than these three, but only these three have been identified in the sampling methodology used in this study.)

The shipyards expressed capacity concerns nine times in total. However, in three cases, multiple shipyards identified the same supplier. For these, only one shipyard indicated a 'high' concern about capacity, while the others rated the same supplier's capacity concern

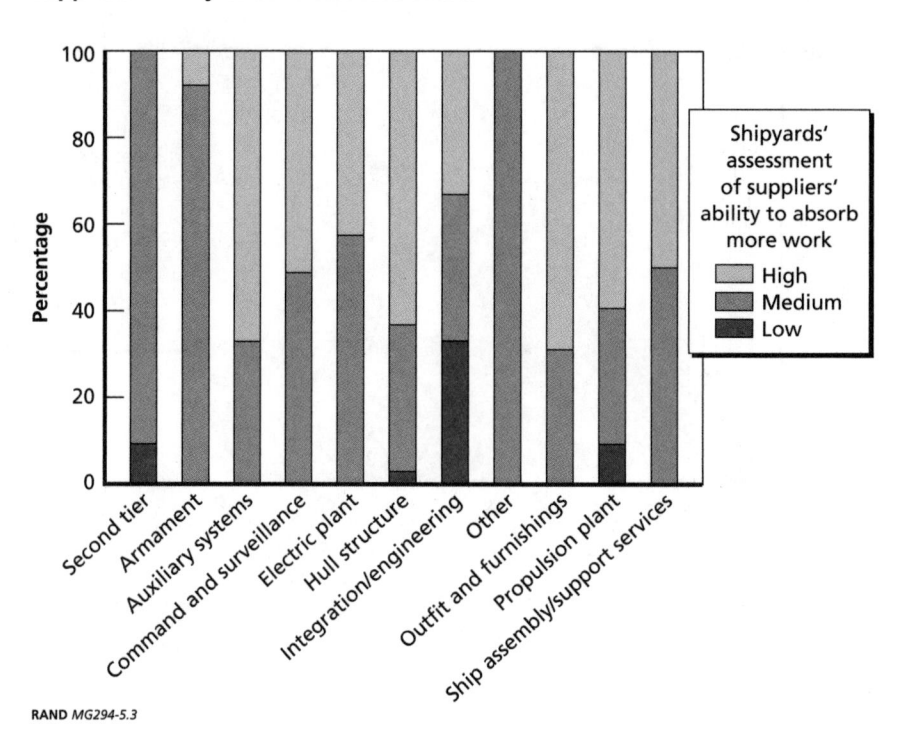

Figure 5.3
Suppliers' Ability to Take on More Work

[Bar chart showing shipyards' assessment of suppliers' ability to absorb more work (High, Medium, Low) across categories: Second tier, Armament, Auxiliary systems, Command and surveillance, Electric plant, Hull structure, Integration/engineering, Other, Outfit and furnishings, Propulsion plant, Ship assembly/support services]

RAND MG294-5.3

as 'medium' or 'low'. One implication is that this is a subjective measure, and the shipyards may not individually have complete insight into their suppliers' operations and ability to take on new work.

The second measure we assessed is the shipyards' concern about the long-term stability and viability of their suppliers. Suppliers rated as 'low' may not be in existence to support future years work. Alternatively, if the rating was 'high', the shipyards have confidence that they (and hence the MOD) will be able to rely on those suppliers in the future.

As Figure 5.4 shows, it appears that the shipyards are relatively confident that most of their suppliers will be viable in the long term.

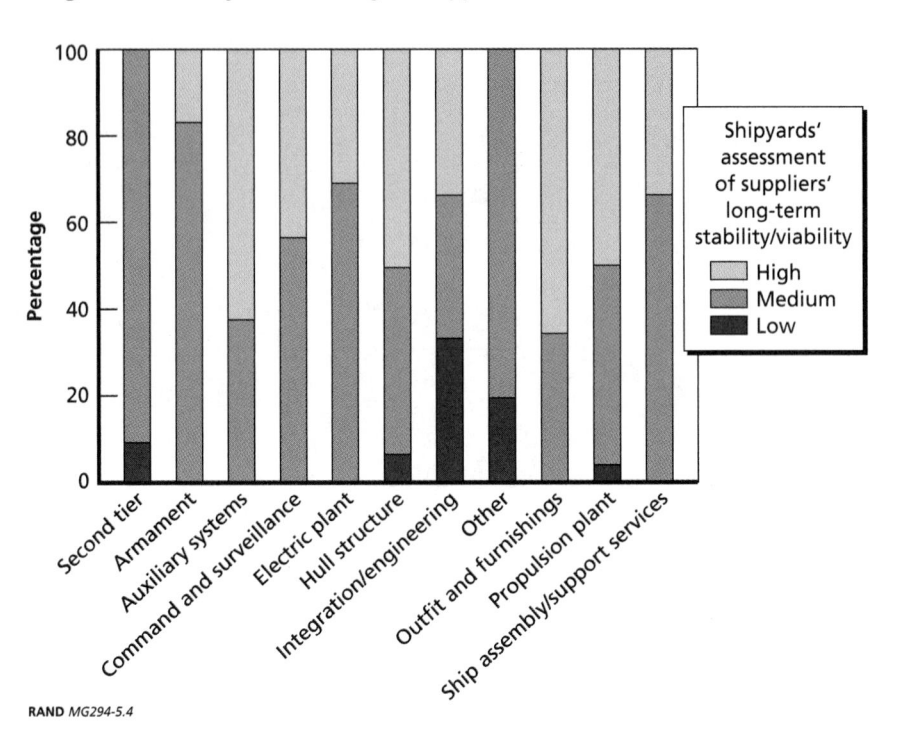

Figure 5.4
Long-Term Stability and Viability of Suppliers

Again, for the one sector that appears most in jeopardy (integration/engineering), the level of risk is the result of one company being of concern out of three total identified for this study by the shipyards.

The shipyards expressed survival concerns about 10 suppliers. In two cases, multiple shipyards identified the same supplier, and for each of these only one of the shipyards suggested that the supplier confidence was at 'low' in terms of its long-term viability.

The third measure refers to how much competition or how many alternatives there are for the goods and services that a particular supplier provides—a measure of dependence on a supplier. However, this indication of risk does not have a simple interpretation. The shipyard may be in a relatively stronger negotiating position compared with its supplier if there are multiple providers for the same good or service. Competition will presumably work to keep the supplier costs low and the quality and service high.

Without competition, the yard may have to pay relatively more. But this possibility would not necessarily make any particular MOD shipbuilding programme more risky. The supplier may be willing and able to take on more work, and it may have strong prospects for long-term survival. Having only one competitor in any particular sector may also be the result of defence consolidation, a natural consequence of the broader downsizing of weapon acquisition programmes.

As Figure 5.5 suggests, there was a bit more concern in the area of competition than for the other two measures of risk. Shipyards indicated that 25 suppliers did not have competition for the goods and services they offered. However, multiple shipyards identified 10

Figure 5.5
Dependence on Supplier/Competition

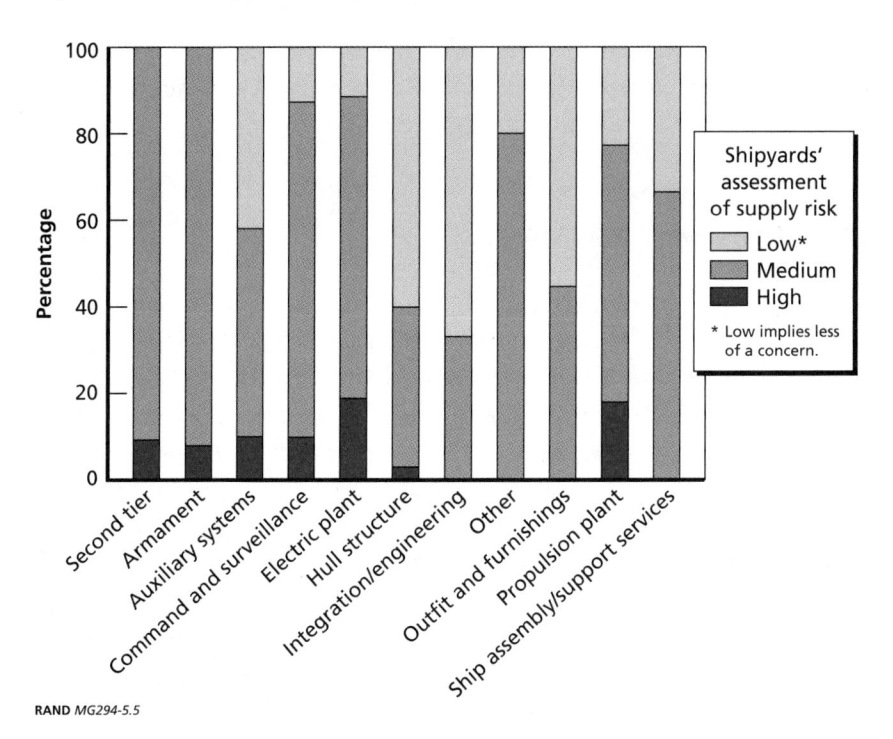

companies as being key suppliers, and in every case there was disagreement about the lack of alternatives. It is possible that the shipyards are buying different goods and services from these suppliers and are providing accurate assessments of the availability of alternatives, but it is also likely that some shipyards have a broader view of the supplier base than others.

Summary

The shipyards buy a wide variety of products from many suppliers, most of which are located in the United Kingdom. There is some concern about the ability of individual suppliers to increase their production, as well as their prospects of long-term survival. These issues could usefully be the focus of further investigation into the supplier base. There are also a number of suppliers for which there may be no competition for the products and services they provide. The defence market in the United Kingdom may not sustain multiple suppliers in these sectors. However, the United Kingdom could assess global alternatives, of which the shipyards may not currently be aware of, to make sure that in worse-case scenarios these particular items could be purchased from nondomestic suppliers.

Supplier Survey Results

Even though we attempted to obtain inputs from all the suppliers for which we had adequate contact information, not every supplier returned the survey. During the follow-up phase, few refused directly, but there were cases in which companies stated that they had 'no-survey' policies. Some promised to return the survey but never delivered, while still others were unreachable and did not return messages. Overall, the response rate was about 21 percent, and therefore, the results of the supplier survey should be viewed as indicative rather than statistically significant. We do not know whether the suppliers that responded are representative of the shipyard suppliers as a whole, but we have no reason to believe that there is any systematic bias.

(Although it may be that suppliers that are most interested in supporting MOD naval work would be more likely to reply, we cannot assert this assumption with any certainty.)

We used the supplier survey to get at measures of supply risk that could not be captured by asking the prime contractors themselves. We asked about the suppliers' dependence on MOD and naval work, on the number of customers and competitors, on their workforce and their recruiting challenges, and on the challenges they face in supporting MOD shipyard programmes.

Demographic Information on Sample Suppliers

A common measure of company size is number of employees. In the sample of companies that responded to our survey, two-thirds were relatively small, with 250 employees or fewer (see Figure 5.6). We report this detail to provide descriptive information on our sample, but we cannot say whether the responses are representative of the shipbuilding supplier base as a whole. (The sampling frame in which

Figure 5.6
Size of Suppliers, by Number of Employees in 2003

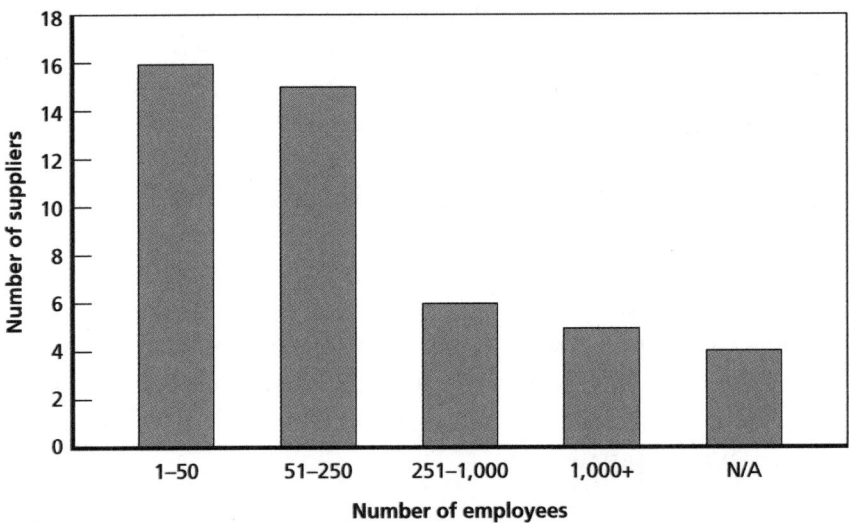

we specifically asked for information on key second-tier suppliers may have led to an overrepresentation of smaller firms.)

Twenty-five of the 48 suppliers that responded to the survey are subsidiaries of larger organisations, and the rest are independent. Thirty-one of the suppliers are privately held, and 17 are publicly owned firms. In terms of country of the suppliers, the distributions roughly followed that of the broader sample shown in Figure 5.2. Of the 48 respondents, 39 were from the United Kingdom, two from Germany, two from Italy, four from the Netherlands, and one from the United States.

Suppliers' Business Base

In the survey of suppliers, we asked the suppliers to divide their sales into six categories, providing the percentage of their total business in each. Broadly, the division was between shipbuilding and other offshore work, and all non-shipbuilding-related work, further divided into MOD, other military, and nonmilitary work. The results from this question can be presented in a number of ways, giving insight into what portion of suppliers' business bases is a result of sales for MOD shipbuilding, for all ship-related work, or for all military work, for example. Figure 5.7 presents the average of all supplier results.

Typically, about half the work of the average supplier is derived from the totality of its shipbuilding, naval, and offshore work, while about a third comes from work that has nothing to do with shipbuilding and does not have a military customer. We also found that the average supplier receives less than a quarter (about 22 percent) of its total revenues from MOD shipbuilding programmes.

This last statistic suggests that most of these suppliers are not overly reliant on MOD naval programmes to sustain their business base and may be able to withstand the unexpected programme changes and production gaps that may be typical of military programmes of all sorts in every country. Examining this one category in more detail provides further insight into suppliers' dependence on MOD shipyard programmes.

Figure 5.7
Average Supplier Dependence on Different Sectors

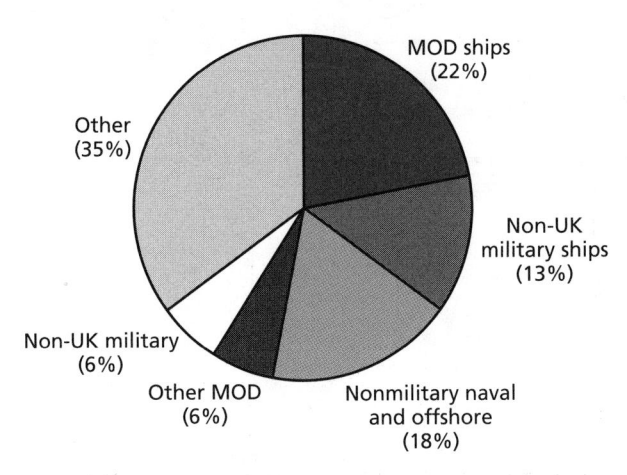

- MOD ships (22%)
- Non-UK military ships (13%)
- Nonmilitary naval and offshore (18%)
- Other MOD (6%)
- Non-UK military (6%)
- Other (35%)

RAND *MG294-5.7*

As Figure 5.8 shows, few suppliers get more than 50 percent of revenues from MOD shipbuilding and repair, and as many as two-thirds derive a quarter or less of their work from this sector. Thus, most suppliers in our sample are not highly dependent on revenues from MOD shipbuilding and repair programmes. One possible implication is that fluctuations in these programmes would have a relatively limited impact on the typical supplier (although clearly some would bear a greater burden).

Even though few companies in the supplier sample are highly dependent on MOD naval work, we see that many are heavily involved in the maritime sector, as depicted in Figure 5.9.

A significant number of the suppliers—in excess of half—derive more than 50 percent of their revenues from shipbuilding, repair, and offshore work as a whole, with one-third of the companies doing more than 75 percent of their work in this sector. This information suggests that the suppliers are likely to have focused on the specialist skills required for shipbuilding, although it should be noted that one-third of the companies do less than 25 percent of their work in the sector.

Figure 5.8
Percentage of Suppliers' Revenue Derived from MOD Ship Programmes

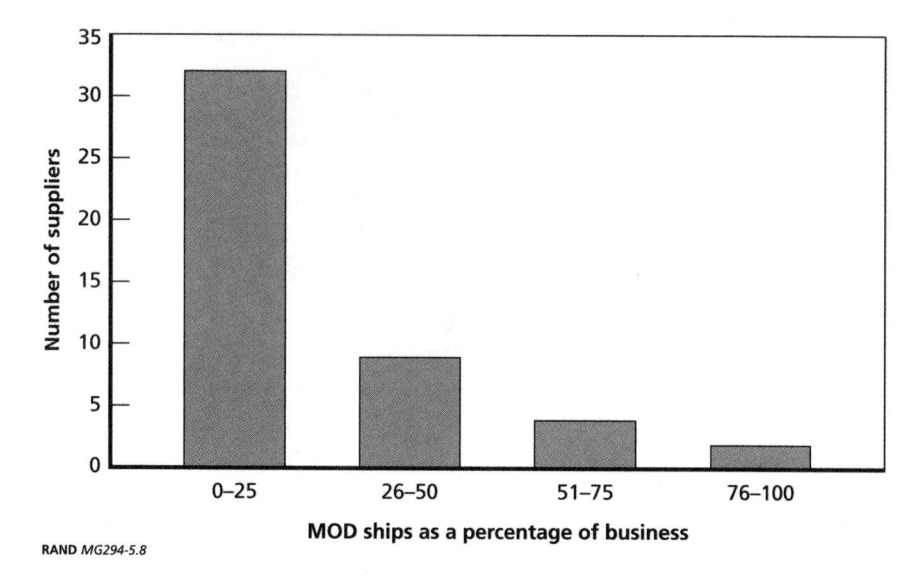

MOD ships as a percentage of business

RAND MG294-5.8

Figure 5.9
Percentage of Suppliers' Revenue Derived from All Ship and Offshore Work

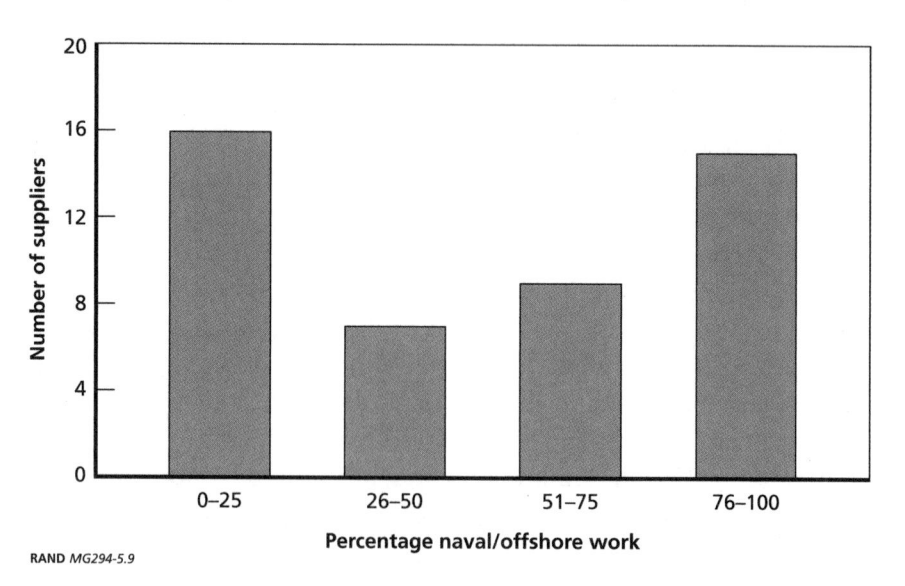

Percentage naval/offshore work

RAND MG294-5.9

Companies that derive a larger percentage of their work from the military may be at risk if budget concerns cause a general draw down in military spending or if the defence climate were to change in other ways.

As Figure 5.10 reveals, the companies we surveyed are fairly evenly divided in terms of their dependence on military work.

Number of Customers

Another measure of supplier strength is the number of major customers that they have. If just a single customer, they are at greater risk for failure should something happen to the prime contractor they support—or in case of a perturbation, to the programme they support. Figure 5.11 lays out the response to the question asking about numbers of customers.

Very few suppliers depend on single customers for either their ship or non-ship-related work. A robust customer base signifies that the suppliers should be better positioned to handle perturbations

Figure 5.10
Percentage of Suppliers' Revenue Derived from Military Work

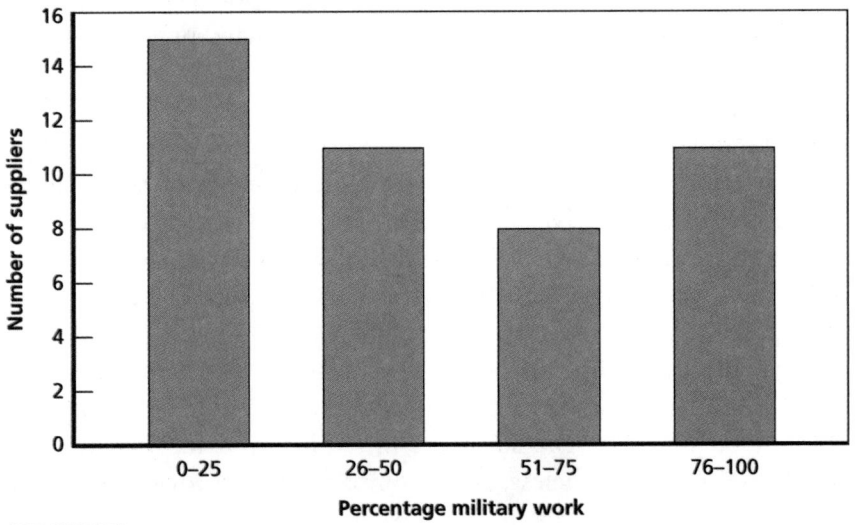

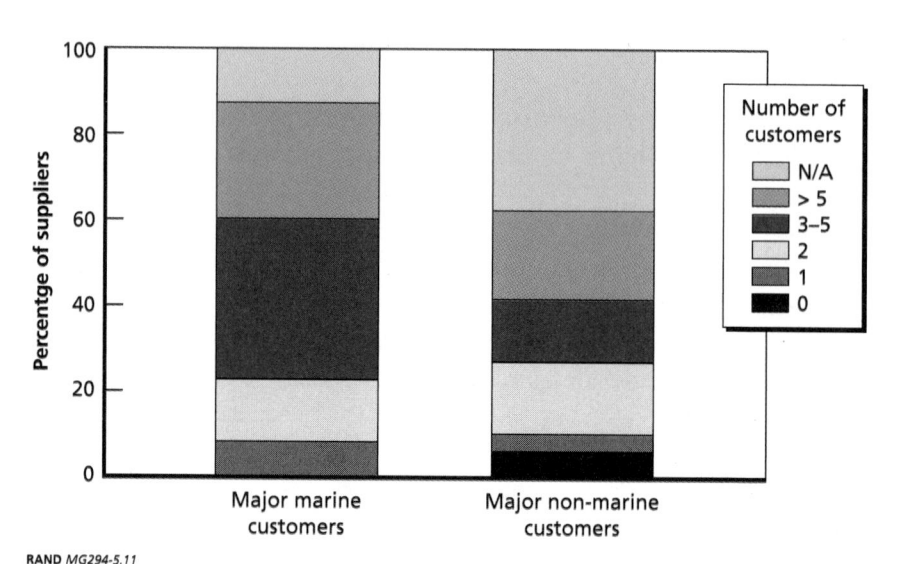

Figure 5.11
Numbers of Marine and Non-Marine Customers

[Stacked bar chart showing percentage of suppliers by number of customers (N/A, >5, 3-5, 2, 1, 0) for Major marine customers and Major non-marine customers]

RAND MG294-5.11

from any single customer. However, all their customers face downturns, which could be problematic.

It should be noted that although the suppliers in this survey listed their major customers, they might not have provided a complete list.

Number of Competitors

One measure of risk that the shipyard survey explored was whether there were competitive alternatives for an existing supplier. We attempted to acquire this information on the supplier side by asking who their most important competitors are. We present these results in Figure 5.12.

There is a significant amount of missing data here. Many suppliers provided no information on their competition. Moreover, we cannot be sure that the suppliers providing information offered a complete list. Nonetheless, most companies that provided informa-

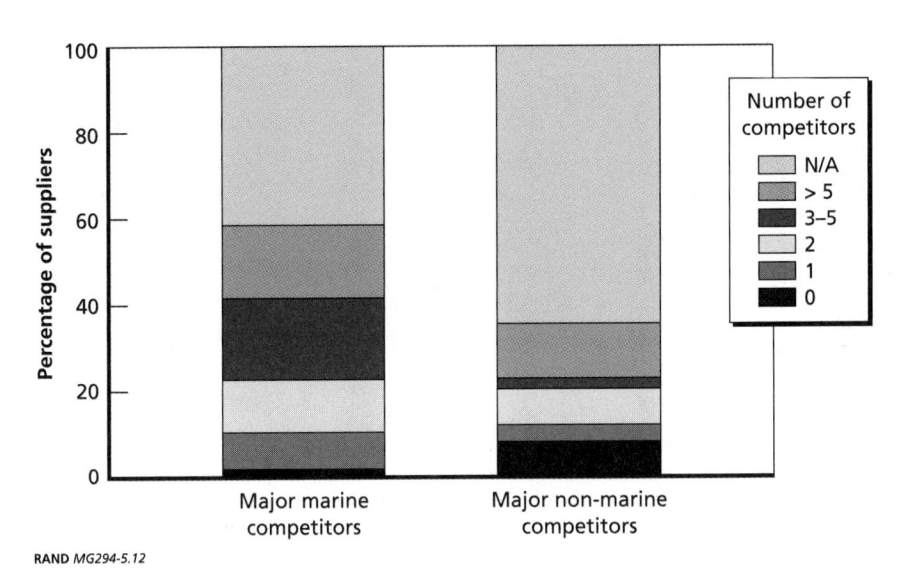

Figure 5.12
Numbers of Marine and Non-Marine Competitors

tion indicated that they did face competition, which may mean less risk for the shipyards, and hence MOD, in these industries (although the competitors may not provide reasonable substitutes for some of what the suppliers in our survey provide).

Recruiting Challenges

We asked suppliers to rate how easy it was for them to recruit in four different labour categories. Although there was a fair amount of missing data, the results presented in Figure 5.13 are still suggestive.

Engineers Presented the Most Challenges for Recruiting

To get at this question another way, we asked the suppliers if they had any particular labour force recruiting challenges. Many suppliers (21) did not answer this question, so they may not have had any problems. Nine others specifically said 'none'. Those that did face challenges identified engineers as hardest to find, with seven men-

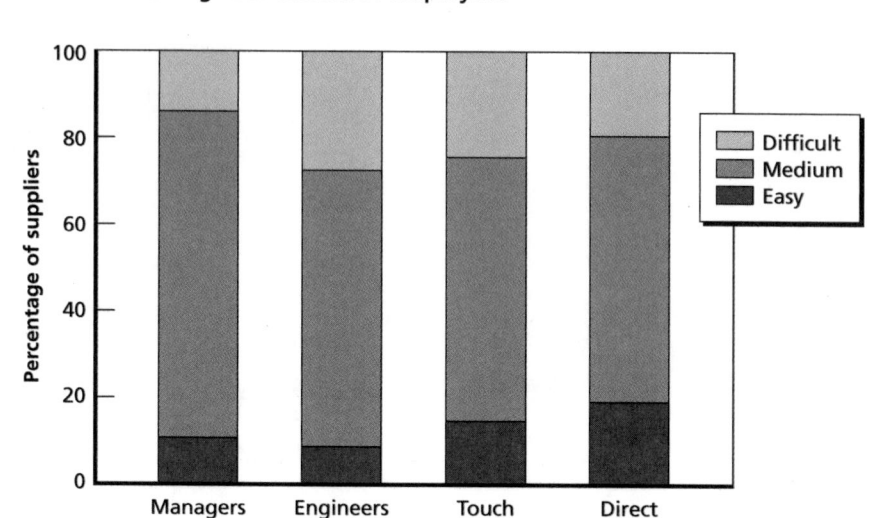

Figure 5.13
Ease of Recruiting Four Classes of Employees

tions. Two companies had problems recruiting employees with the ability to create three-dimensional CAD drawings, and two had problems finding skilled machinists. The rest of the skills that were mentioned as problematic received only one mention each, including project managers, sales people with appropriate background, electricians, metallurgists, ultrasonic tube testing technicians, computer numerical control operators, fabricator welders, and job-specific production staff. These results indicate that the suppliers face similar challenges as the shipyards in terms of specific recruiting needs (see Chapter Three).

Challenges Working for the MOD

We also asked the suppliers directly about the challenges that would limit their ability to meet the MOD's ship acquisition plans. Fourteen expressed no concern. For the others, the most common concern related to the fact that doing business with the MOD involved a lot of unknown factors regarding the MOD's plans. Sixteen suppliers indicated some aspect of MOD's planning and budgeting process as the biggest challenge. However, it is clear that there is interest in MOD work, since seven suppliers focused on winning the business as their main issue. The other main concern was getting the right kind of employees needed to do the work.

The following comments are typical concerns taken directly from the surveys:

- 'Contracts placed at same time with short delivery schedule'
- 'Ease the feast to famine periods of high and low demand'
- 'Lack of consistency in planning'
- 'Lack of continuity in MOD ordering programme'
- 'Lack of continuity and decisionmaking by MoD'
- 'Not being given specific early information on requirements'
- 'Programme delays'
- 'Uncertainty over procurement programmes'
- 'Insufficient lead time'
- 'There is little point in trying to plan because the projects are so undefined and there is very little if any guarantee'.

Concern or complaints about serving customers in any industry may not be uncommon. However, as the MOD examines what it can do to strengthen the shipbuilding industrial base, it should be aware that if suppliers could plan operations based on more consistent programmes, they may become stronger and more willing to operate in the naval shipyard sector. One of the risks that should be assessed when MOD alters its shipbuilding plans (e.g., by extending programmes) is the effect of changes on the supplier base, especially those at higher risk of leaving particular lines of business.[3]

---

[3] This risk has not gone unrecognised. 'He [Murray Easton, BAE Systems Submarine's managing director] is also conscious that delays in the Astute programme have put parts of the submarine industrial base under strain' ('Astute Sets Out on the Long Road to Recovery', 2003, p. 30).

Summary

Although our response rate of 21 percent does not allow us to assert that these results are statistically significant, we believe that they are suggestive of the larger shipyard supplier industrial base. The companies that responded to the supplier survey were a variety of sizes and were divided between independent, stand-alone firms and subsidiaries of other companies, as well as between private and public firms. They did not appear to be overly dependent on MOD shipbuilding programmes for their business, although many did a majority of their business in various aspects of shipbuilding and offshore work. It was also typical for firms to have both multiple customers, putting them less at risk if something happens to one of their customers, and multiple competitors, meaning that their customers would have alternatives if something were to happen to their suppliers. Finding appropriately trained engineers is their biggest recruiting challenge, and the perturbations and uncertainties in MOD shipbuilding programmes is their biggest difficulty in supporting this work.

Results from Linking Shipyard and Supplier Surveys

We attempted to match information from the shipyard and the supplier surveys to see if there were any patterns that might indicate areas of supplier risk about which the MOD should be concerned.

Of the 48 suppliers that responded to the survey, six were identified as being high risk in one of the three categories on the shipyard survey. We explored the extent to which their business base was devoted to MOD shipyard work and found that the numbers varied from 15 to 30 percent. Three of these suppliers are subsidiaries of other firms, so they might be supported in economic downturns. Alternatively, these subsidiaries could be sold off or shut down if they did not meet certain profit targets. Finally, only one company that was rated as having few alternatives or limited competition by its shipyard customers provided information on its own competition, which it rated as being 'numerous'.

It is difficult to reach any conclusions about the health or strength of the suppliers based on this matching of the two surveys. Nevertheless, it does suggest the possibility of strengths in the supplier base resulting from a broader customer base than their shipyard customers may be aware of. (However, it is also clear from reports in the press and other sources that some suppliers identified as problematic by the shipyards are very much at risk.)

Developing an Effective Supplier Strategy[4]

The question of how to best manage suppliers is by no means a new one. Over the years, a considerable amount of literature has emerged describing the best practices that commercial firms use to manage their supplier base.[5] Although this research is too broad and diverse to review here, there are some consistent themes we can touch on.

Researchers in the field recommend taking a proactive approach to managing the supply base. Suppliers play an important role in the production of any final product and should be managed with their strategic role in mind. Steps for firms to follow are (1) conduct a firmwide spend analysis on all money spent on suppliers; (2) rationalise the supply base, consolidating contracts where possible (and where this fits the legal requirements for competition); (3) establish long-term partnerships with the best, most strategically important suppliers; (4) help key suppliers improve quality, cost, and service; and (5) integrate key suppliers into the organisation.[6]

---

[4] This section is meant to be an introduction to some best-practice supplier management information to the general reader, not a critique of current MOD or shipyard practices in this area, which were not a subject of investigation in this study.

[5] For example, Monczka, Trent, and Handfield (2002), Cavinato and Kauffman (1999), Gattorna (1998), and Laseter (1998). There are too many books and articles on this subject to offer a complete list here.

[6] Moore et al. (2002). This report also documents specific cost saving and performance improvements that have resulted from strategic supply chain management.

Part of the strategy is to 'segment' the supply base. A number of researchers[7] recommend dividing suppliers into categories based on multiple dimensions, for example, viewing the total amount spent at suppliers in light of strategic importance or risk. If the supplier base is categorised based on scoring 'low' or 'high' in these two dimensions, an appropriate strategy can be identified for each of the four categories that result. Low risk/importance and low spend items can be consolidated into larger contracts and issued on the basis of cost and reliability of the supplier. It is more critical for companies to carefully manage suppliers that are high risk/importance and high spend, perhaps partnering with them to work on reducing costs and increasing quality. In such instances, long-term contracts might give the suppliers reason to invest in the relationship with the customer.

Furthermore, best-practice customers not only manage their own suppliers but also try to have their first-tier suppliers manage the second-tier suppliers in a similar fashion, with the goal of taking costs out of and improving quality in the entire supply chain. The MOD could certainly encourage its own suppliers—the shipyards—to invest in best-practice supplier management techniques and could also use performance in this area as one of its assessment criteria.[8]

Conclusion

The methodology we followed did not reveal a major problem in the ability of the supplier base to support MOD shipbuilding programmes. However, there are a few suppliers and sectors that may present some risk. The MOD's SRG could follow up on these specific companies. Our methodology also did not include government-furnished equipment that is provided by the MOD directly to the shipyards.

---

[7] Two of the first were Bensaou (1999) and Tang (1999).

[8] This may already be in place; again, we did not research MOD practices in this study.

However, the absence of a systematic problem does not mean that the MOD would not benefit from proactive supplier management strategies, both of its own prime contractors and from the prime contractors' management of their own supply chains. There is considerable evidence of reductions in cost and improvements in service and quality resulting from a more considered and strategic approach to supplier management.

CHAPTER SIX
Nontraditional Sources for Naval Shipbuilding: Commercial Shipbuilding and Offshore Industries

The previous chapters have focused primarily, though not exclusively, on the ability of the naval shipbuilders and ship repairers to meet the MOD's plans over the next 15 years. Given the large peak demands for both labour and facilities, the MOD will need to explore many possible options to either mitigate or overcome these peaks. Options we have examined so far are how level-loading, outsourcing, work-sharing, and workforce recruiting may or may not help to fully mitigate these peaks. Another option for the MOD is to engage a broader set of firms for the design and production process.

The United Kingdom, at one point in time, had fairly extensive industries in both commercial shipbuilding and offshore fabrication. As a result of either a decline in business or competition from foreign sources, these industries are either vastly reduced from their peak or are underutilised at present. Given that these industries share a common market segment—maritime industries—the natural questions that arise are whether and how these industries can contribute to the MOD shipbuilding plan. In this chapter, we will explore these issues. Given the limited information from some of these firms,[1] the discussion that follows will be largely qualitative.

---

[1] Some firms did participate in the study, whereas others either declined to participate, did not want to provide sensitive information, or did not respond.

Declining Markets for Offshore and Commercial Work

At one time, the United Kingdom was the largest commercial shipbuilding nation in the world.[2] This position declined over the 20th century, with the country providing only a small portion of the total commercial output.[3] The reasons for this decline are too involved to address in this document, and there are excellent treatments of this subject elsewhere.[4] To illustrate the recent decline in UK commercial shipbuilding, let us examine some output data presented in the recent RAND study.[5] Figure 6.1, taken from that study, shows the gross registered tonnage delivered each year since 1980.[6] We have abstracted the data to show only deliveries to commercial customers. As can be seen, the overall output today is a fraction of its peak in the early 1980s. As a result, there are very few commercial shipbuilders active in the United Kingdom. In fact, many surviving commercial shipyards focus on repair work.

Similarly, the workload for the offshore industry also has declined. Figure 6.2 displays the decline in new production workload for the North Sea sector based on data from KBR. As shown, the current new production work is a quarter to a third of what is was in the mid-1990s.

We provide these two figures not to define the capacities of the industries but rather to show that the UK business for both sectors has declined significantly in recent years. Thus, one could infer that there might be available capability that could be applied to naval shipbuilding. In fact, some firms (such as KBR and AMEC) are actively pursuing such opportunities. The question is, what is the extent of that capacity?

---

[2] Johnman and Murphy (2002); Burton (1994).

[3] Birkler et al. (forthcoming).

[4] Johnman and Murphy (2002); Jamieson (2003); Burton (1994); Winklareth (2000); Owen (1999); Walker (2001).

[5] Birkler et al. (forthcoming).

[6] The data do not include vessels of less than 100 gross tons.

Nontraditional Sources for Naval Shipbuilding   139

Figure 6.1
Decline of Commercial Shipbuilding in the United Kingdom Over the Past Two Decades

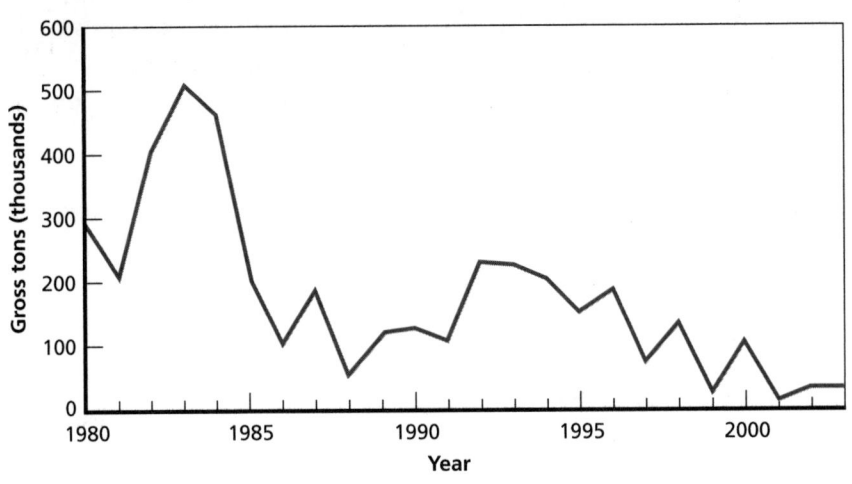

RAND MG294-6.1

Figure 6.2
UK Offshore New-Build Market

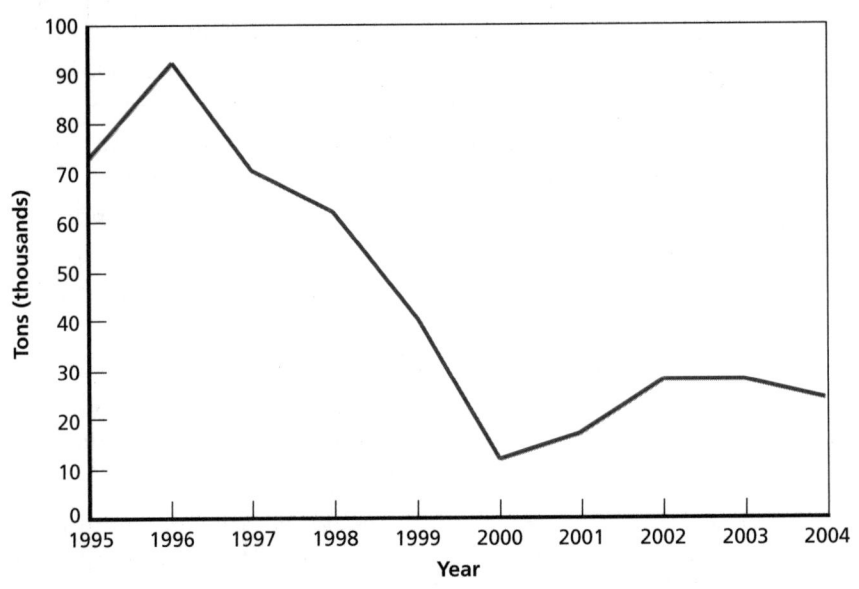

RAND MG294-6.2

Potential Resources Available

To gauge the capacity of the aforementioned industries to take on naval shipbuilding work in the United Kingdom, we contacted a number of firms that fit into the broader maritime sector that either (a) supplied data to this study, or (b) interacted with RAND on previous studies:

- A&P Group
- AMEC
- Appledore (now part of DML)
- Atkins Global
- BMT
- Ferguson Shipbuilders
- KBR Caledonia
- Harland and Wolff.

We cannot make a clear distinction between exclusively commercial and exclusively offshore resources. Many of these firms (e.g., Atkins Global, AMEC, KBR, and BMT) provide design, engineering, management, and/or production services to a broad range of clients and industries (commercial, military, and offshore). Similarly, the medium-sized shipbuilders (Harland and Wolff, Appledore, and Ferguson Shipbuilders) have engaged in a combination of commercial and naval production. By including these firms in this chapter, we are not implying that they focus exclusively on any particular industrial sector; rather, the combination was done to simplify presentation and to show what resources might be available from firms that work in other areas of the maritime sector.[7]

We have included Ferguson and BMT data in this sample, even though they are included as part of the labour supply analysis of Chapter Three. We include them again to show the magnitude of the

---

[7] Likewise, Swan Hunter has engaged in projects for the offshore industry. DML and VT have each built private yachts. We do not include these firms here because they have been included in previous chapters.

resource available from additional suppliers and to protect other firms' data.

Labour

As mentioned earlier, we do not have sufficient data from these firms to conduct a detailed demand and supply analysis similar to the one we did for the naval shipbuilders and repairers in Chapters Two and Three. However, it is useful to describe the magnitude of the potential labour resources that might be available at these firms. We will summarise the labour currently engaged, at one time engaged, or thought to be readily available.

There are two qualifications to note about the data from these firms:

1. The labour resource data from KBR Caledonia, Atkins Global, and AMEC includes resources that would be available beyond those available in the United Kingdom. Furthermore, some of these firms have noted that significantly greater resources are available worldwide through their parent organisations. For example, a firm might have included an overseas engineering office in its workforce data. Obviously, there would be security and political concerns that must be considered before leveraging such resources.
2. The data for Harland and Wolff, A&P Group, and Appledore were obtained circa 2001 in a previous RAND study for the MOD. The data are representative only, given that the business situation of the firms might be quite different today.

Table 6.1 summarises the labour data by the five categories introduced in Chapter Two. The table contains an additional 'unspecified' category into which we placed employees we could not allocate into one of the five categories. The numbers have been rounded. In the medium-sized shipbuilders, we include Appledore, Ferguson, and Harland and Wolff. The other firms are collected in the 'other firms' category.

Table 6.1
Labour Resources of Medium-Sized Shipbuilders and Other Firms (number of employees)

Skill	Medium-Sized Shipbuilders (number of employees)	Other Firms (number of employees)
Management	130	2,300
Technical	330	17,000
Structure	550	700
Outfitting	140	820
Support	130	690
Unspecified	—	1,000

Strikingly obvious from the table is that there is a large number of design and technical resources employed by these firms. Although not all this technical resource has naval expertise, it is a resource that could be used to supplement design teams and management on naval production programmes. Not to be ignored are the medium-sized yards, which too can bring resources for module block construction or even smaller vessel fabrication. For example, Appledore constructed the two *Echo*-class survey vessels under subcontract to VT, and Ferguson Shipbuilders has constructed fishery protection vessels for the government. We must caution the reader, however, not to expect that all these resources would be at the MOD's disposal or that they even exist currently. The values represent *potential* workforce. In some cases, the naval shipyards are considering working with these firms as a way to outsource peak workload.

An important characteristic of 'other firm' workforce worth considering is that it tends to be quite mobile. In other words, it is not uncommon for the workers to move to different sites, depending on where the work is located. In essence, they move to the work; the work doesn't come to them. Their movement may be to other countries, as well. It is not uncommon to have a significant portion of the workforce be expatriates. To illustrate this point, a study of offshore workforce skills in Nova Scotia examined workers' willingness to move to other locations for jobs.[8] The authors found that 40 percent of the workers surveyed were willing to travel outside Canada for work. Only 16 percent of those surveyed said they were willing to travel no more than 100 kilometres. Such mobility greatly aids the offshore industry to level demand and maintain an appropriately sized and skilled labour force.

Facilities

The commercial and maritime sectors also have facilities that could be used for naval production (if not engaged for other work). In fact, some of the traditional naval shipyards are considering or are already using some of these facilities. For example, Swan Hunter is in the process of reactivating a former offshore fabrication facility for use in naval shipbuilding programmes. KBR Caledonia has a fabrication and assembly facility in Nigg with one of the larger dry docks in the United Kingdom. Likewise, Harland and Wolff has a large dry dock with associated facilities, which have been used for both naval and commercial shipbuilding in the past. BAE has considered using the Inchgreen dock in Greenock for future production. Ferguson and Appledore have facilities capable of producing smaller naval vessels (approximately up to frigate-sized ship). We have summarised the facilities capacity for these firms in Chapter Four.

Strengths and Weaknesses of Using Offshore Firms in Naval Production[9]

The offshore industry has traditionally produced commodities more akin to structures and petroleum process facilities.[10] So a natural question arises: How might the offshore industry contribute to naval

---

[8] Chaundy (2002).

[9] These statements are not specific to any one firm but are generalised for the industry. Any individual firm might have different characteristics.

[10] Recent production of floating production storage and offloading vessels has blurred the line between a vessel and production facility.

shipbuilding? The industry could bring several strengths to shipbuilding. It has experience working on collaborative programmes and alliance-type structures and has a network of suppliers and subcontractors that may be untapped by the naval shipbuilders. This subcontractor resource could be used to increase the outsourcing done by the shipyards. The offshore industry is experienced with distributed design (engineering done in more than one location) and has broad programme management and logistic skills on collaborative programmes. Further, modular construction is the norm for the industry.

That said, the offshore industry does have some weakness with respect to naval shipbuilding. It lacks specific domain knowledge of military ships and therefore will be unfamiliar with the systems complexity (particularly weapon systems), the design and build standards, and the testing and commissioning requirements. Further, offshore firms will generally (although not all) have limited experience with MOD on new systems acquisition. Therefore, procedures, policies, and general programmes issues will need to be assimilated. Lastly, not all offshore firms have extensive steel plate/steel structure panel-line fabrication facilities. Thus, the offshore industry, as a whole, will not serve as an extensive resource for structural work.

The strengths of the offshore industry play well to the need for the naval shipbuilding industry to work in a more collaborative fashion. As we have stated earlier, better work-sharing between the shipyards will be necessary to meet the peak labour demand. The offshore industry might be the catalyst to make this work-sharing happen and function successfully. It is not likely, however, that the industry will feature strongly in direct fabrication; it might feature more prominently in assembly and integration.

As with any owner-contractor relationship, an owner that is able to well define a programme up-front and minimise late changes achieves better performance. A study of exploration and production (mostly offshore) programme performance over the past decade found that, on average, offshore programmes experienced approximately 10 percent cost and schedule growth over their authorisation plans.[11] About one-eighth of the programmes were termed 'disasters' and had an average of about 30 percent cost growth and 35 percent schedule slip. For projects that were well defined and that controlled changes, cost growth was approximately *negative* 5 percent and schedule slippage was less than 5 percent. The MOD's expectations of performance will have to be tempered by the fact that it has a tendency to have higher levels of change on its programmes than does a commercial customer.[12] The MOD should not expect commercial levels of performance on programmes in which there is a high level of change. Similarly, commercial firms will need to expect greater levels of change and adjust their control processes accordingly.

Finally, it is important to note the differences between the products that each industry produces. As described in Chapter One, warships are complex systems integrating many functions (e.g., communications, food service, weapon systems). The offshore industry produces a variety of products, from drill, support, and pipe laying ships to offshore production platforms. Although these ships have similarities to other commercial vessels, they are quite different from combatants in terms of system complexity and the density of systems. Offshore platforms are even more different. These products can be viewed more as structures with processing equipment and support facilities on topsides. The topside plant is typically produced as skidded units or modules (often produced at several locations) with much of the equipment and outfit (piping and electrical) already installed. Therefore, the final outfitting activities on the platform are minimised. This contrasts with combatants, where weapon systems and sensors are installed at the end of the production process to take advantage of the most advanced equipment. So, while some of the offshore products are similar to commercial maritime ones, they are quite different from a combatant.

---

[11] Merrow (2003).

[12] Arena et al. (forthcoming).

Summary

The UK commercial shipbuilding and offshore industry have resources that may help to produce ships for the MOD. These resources are currently underemployed and could be available to the MOD for its shipbuilding programme. The medium-sized shipbuilders have played and could play a role in the production of ships—from a builder of blocks and modules to a producer of smaller vessels. The offshore industry also has potential facilities and labour resources to contribute to the programme. However, its role, if any, will need to be carefully matched to its capabilities and skills. For programmes that are similar to commercial products, the industry could play a broader role in the management, design, and production. However, for combatant ships, their role will be more limited.

CHAPTER SEVEN
Issues for the Ministry of Defence to Consider

Summary

The MOD will be embarking on an ambitious shipbuilding plan to renew the naval fleet over the next several years. As we have seen in the previous chapters, the MOD will face a number of capacity issues in implementing this plan. The crux of the capacity issue is the number of simultaneous shipbuilding programmes that will occur between 2007 and 2012. This overlap of programmes will cause a large spike in the demand for resources. The resources that we explored are labour, facilities, and suppliers.

Labour Demand

The peak in the production labour demand at the shipyards and firms is a result of the overlap of four large programmes: Type 45, CVF, MARS, and Astute. The peak labour increase for the direct workforce is about 30 percent greater than the current employment levels of those firms that participated in the study. Table 7.1 shows the relative growth in each of the skill areas examined.

The most significant increase (in absolute sense) is in the structural and outfitting trades. Once past the peak, overall workload demand steadily declines for the foreseeable future. Although not shown in this report because of the business sensitivity, a few naval shipyards face workload gaps or erratic demands despite the overall high aggregate level of demand.

Table 7.1
Percentage Growth at Peak Demand Relative to Current Employment Levels (direct workforce only)

Skill	Percentage Growth
Management	10
Outfitting	42
Structural	48
Support	28
Technical	21
Total (weighted average)	32

The technical workforce demand presents a more difficult challenge. We showed that there is a sharp drop-off in demand for technical workforce in the near term, which results mainly from the rundown of the design work for the Type 45 and Astute. Later, the trend reverses dramatically as the CVF, MARS, and JCTS place near simultaneous demand for technical workers. In the span of a few years, the demand for technical workers nearly doubles from its low.

Variation in the technical workforce is much more problematic for the firms. First, these technical workers take years to develop and train. So, new hires are limited to what higher education produces. Furthermore, technical workers have general skills that allow them to move to other industries. Therefore, after a period of reduced demand, technical workers who are made redundant will likely leave the industry.

The statements summarised above regarding the demand for production and technical workforce reflect a scenario that is based on the MOD's current plans. We did, however, explore several alternate acquisition scenarios. In almost all cases, regardless of programme variation, the potential MOD plans result in a significant increase in the amount of production labour demand. This increase in labour demand will force the shipbuilding industrial base to rapidly increase its workforce, especially in specific outfitting, structural, and technical skills. After the period of peak labour demand, the amount of direct workers needed to build the future MOD ships will decrease, and will continue to decrease into the foreseeable future. The demands for the technical workforce under the alternative scenarios are much the same as the current plan. After an initial decline, the demand for technical workers increases drastically. One notable exception is the scenario in which we included a Future Submarine (with a new design). For this particular scenario, there are quite substantial demands for technical workers past 2010.

One potential way for the MOD to reduce these peak demands is to employ some level-loading strategies. These strategies entail shifting programmes later in time, extending build schedules, and increasing build intervals. For one specific scenario in which we employed these strategies, the peak workload was reduced. However, the MOD will need to consider operational needs in determining whether such an approach is feasible.

Labour Supply

The shipbuilding and repair industry will be challenged to meet the peak workforce demands as outlined above. As many in the industry have observed, the naval shipyard workforce is ageing and will significantly decline in the next few decades. Although there is no current shortage of workers, the shipyards expressed concern about the future availability to recruit particular skills (e.g., design, electrical, test and commissioning, and steel workers). Many of the shipyards have begun apprentice programmes in recognition of the ageing problem. These programmes are aimed at maintaining current/core workforce levels and not necessarily to meet future peak workload.

There are, of course, other labour sources from which the shipyards can draw workers. For example, some of the shipyards have recently made workers redundant. There is the potential to rehire these former shipyard workers. There is also the opportunity to draw workers from related industries and from among the general unemployed. Another alternative for the shipyards is to rely more heavily on outsourced activities—a trend that has been increasing as of late.

Despite these additional sources of labour, we feel that it will be difficult for the shipyards to grow to meet peak labour demands. It takes time and effort to find and hire new workers. Further, new workers to the industry take time to train and require at least two to three years to become moderately productive. Therefore, if a shipyard hires too rapidly, it will have a hard time absorbing new workers and maintaining productivity. Assuming a modest growth rate, the shipyards as a whole may not be able to meet peak labour demand for production workers. Even under the level-loaded scenario, the shipyards will be able to approximately meet the peak production demands. For the technical workforce, there are currently enough workers at the firms to grow to the needed peak levels, but only if these workers are retained through the near-term downturn. In all, meeting the peaks in workload demand will require that shipyards share work to a greater extent than they do now (although programmes such as Type 45 should be indicative of the feasibility).

Facilities

A variety of facilities are required for the production of naval ships. We discussed the facility implications of the current plan in Chapter Four, where we concentrated on FA facilities (e.g., docks, slipways, and land-level areas) and AO locations (mainly piers and quays). This choice was mainly a result of data limitations and inconsistent measures for other facilities, such as shops and manufacturing halls. We determined that there are a number of challenges that the shipyards face with respect to facilities. For the Type 45 programme, there will be a substantial demand for FA and AO locations because of the build interval between ships (assumed to be six months). Although the facilities on the Clyde might be able to handle this schedule (with facilities upgrades and some careful scheduling), extending the build interval of the Type 45 to nine months might help to alleviate any potential problems and make the build schedule more robust.

The CVF and MARS programmes face challenges because of the sheer size of the ships. CVF assembly will require some facilities upgrades and investments to be made somewhere in the United Kingdom. There is no assembly location in use that can handle the CVF ships without modification or upgrade. A further complication for the CVF vessel is whether the FA location is also used to build portions (large blocks) of the ship. There is a potential overlap between the assembly of the first hull and the production of blocks for the second hull. This implies that either the second hull's blocks will need to start construction outside the assembly dock or be delayed until the first hull leaves the dock.

Similarly, the larger MARS ships envisioned (assuming a Panamax-sized ship) will equally be challenged for an FA location. Although there are facilities in the United Kingdom that could construct these ships, it is likely that at least two facilities (or a facility that can construct two ships at once) will be needed based on the notional build schedule. In most cases, any of these candidate facilities will need to be upgraded or reopened—thus requiring investment.

Suppliers

As stated earlier, suppliers provide a substantial portion of the total value of the ship. Our survey of both the shipyards and the suppliers indicates that there will be generally few issues with the increased workload for the suppliers. For the most part, the suppliers do not rely on MOD business, so they are less subject to the variations in demand (in contrast with the shipyards). Further, most of the suppliers are based in the United Kingdom. However, the suppliers have indicated that the uncertainty in the MOD's programme hinders their ability to plan and invest in a timely manner.

Potential Remedial Actions That MOD Can Take

The current shipbuilding plan of the MOD will be challenging. Without any remedial action, the MOD could face schedule slippage and cost increase for its ship programmes. Although we do not believe the current plan is impossible, the MOD could implement a number of actions to make the plan more robust. The following are some short-term considerations:

- **Consider ways to level-load the labour demand.** We have addressed this mitigation approach in Chapter Two. In essence, the MOD will need to carefully consider the timings of various programmes. Some programmes will need to be shifted later, while others may need to have increased build intervals. We have shown an example of the effect of these level-loading approaches at the aggregate labour level (total industry), but some care must be taken such that a particular shipyard is not adversely affected. Any levelling plan must be made in conjunction with industry with an eye towards the effect on individual firms. One major advantage of the level-loading approach is that it tends to level funding requirements.
- **Work with the DTI (and other government agencies) to encourage training in skills that have competing demands outside the shipbuilding industry.** There are a number of trades and skills that the shipyards recognise as being difficult to recruit or hire. To meet peak workload, the shipyards will have to hire and train new workers. However, after the peak, workers will likely become redundant. Therefore, the UK industry should focus on training skills that are readily employable outside the shipbuilding industry. In this way, any resulting unemployment can be minimised. There are a few skills/trades that are readily employable but are still sought after in the shipbuilding industry. These are electricians, engineers, and drafters. The MOD should discuss with the DTI (and other government agencies) the potential of training programmes or incentives to cultivate these skills.
- **Consider relaxing the shipbuilding industrial policy to mitigate problems resulting from peak demand.** For trades and skills that are shipyard specific, such as structural workers, it makes little sense to hire and train workers to meet the peak demand only to have to make them redundant a few years later. The MOD should reexamine its industrial policy with respect to obtaining work content overseas. For the structural example, the policy might allow UK shipyards to obtain major units or sub-assemblies from abroad in cases in which there is peak demand

and it is not possible to easily obtain the content domestically. We are not suggesting, however, procuring the entire ship abroad but merely the work that can reduce transient peaks in the industry.

- **Encourage the use of more outsourcing.** One way that commercial shipbuilders manage variable workloads is to employ outsourcing vendors that provide services and goods. These vendors are able to better level workload as they work across an entire industry (and perhaps several industries). Examples of these services are painting, electrical power distribution, joinery, HVAC, and combat systems. The use and benefits of outsourcing has been a subject of a recent RAND report,[1] which details how the use of outsourcing varied quite substantially between the UK shipyards and was generally low compared with other European shipyards. With the exception of Swan Hunter, the UK shipbuilders have not relied heavily on outsourcing. This trend is starting to change. For the Type 45, the cabins will be procured as modular units from a supplier. The MOD should encourage, as much as possible, any outsourcing efforts by the shipyards.

- **Evaluate the future of shipbuilding in Barrow.** With the current realignment within BAE Systems, the Barrow-in-Furness facility is exclusively dedicated to submarine production. However, the shipyard has produced many surface ships, including the most recent LPD(R) class. The end of surface ship building in Barrow resulted in significant redundancies and the closure of some facilities. Barrow remains an untapped source of production capability and could likely play a significant role in the coming shipbuilding programme. The MOD and BAE need to evaluate what, if any, future makes sense for Barrow regarding surface shipbuilding.[2]

---

[1] Schank et al. (forthcoming).

[2] Recently, BAE has changed its strategy with respect to surface shipbuilding at Barrow and is now seeking opportunities for the facility.

- **The medium-sized shipyards can help to meet some of this demand peak.** Shipyards such as Ferguson, Harland and Wolff, and Appledore (if DML chooses to employ it for military rather than commercial fabrication) can play a role in meeting the peak demands. These shipyards can build blocks or structural units to ease the demand for structural workers. These medium-sized yards might be able to produce smaller ships, as they have done for survey vessels.
- **Explore the utilisation of facilities for Type 45, CVF, and MARS.** Our high-level analysis indicates that there may be facility challenges for these programmes. The MOD needs to understand where there are potential conflicts and the actions that can be taken to mitigate them.
- **Have the SRG investigate the suppliers that are thought to be at risk.** In our surveys, the shipyards identified certain suppliers they felt to be at risk. It might be worthwhile for the SRG to interact with the shipyards and suppliers to better understand the ones at risk and any corrective actions required.

Beyond these immediate actions, there are some other longer-term issues that the MOD needs to consider.

The MOD Needs to Make Long-Term Industrial Planning Part of the Acquisition Process

Long-term industrial planning must become part of the process that the MOD uses to define the timing for the various naval requirements. Being able to understand the implications of various programme timings on the shipyards will allow the MOD to have more constructive dialogues with the shipbuilding industry concerning timings and workloads. It will also allow the MOD to quickly screen out infeasible plans and seek alternatives. Such planning should be recognised as part of being a 'smart customer'.

Currently, each IPT is responsible for defining its own industrial plan. These activities are primarily done in isolation from the other IPTs. So, each IPT attempts to choose the best plan with respect to its own programme. Yet, the resulting overall DPA position might be suboptimal. This suboptimisation arises from the fact that programmes that are early in the process (ones without a defined build policy) must make due with the industry availability that results from the other IPTs' choices. The shipyards are viewed as first-come, first-served resources, or it is industry's problem to sort out the potential conflicting demands. Thus, there may be cases that arise in which a particular programme has limited choices in terms of available facilities, may have to engage in a more risky build strategy to meet the schedule requirement, or finds that a necessary resource is unavailable. There is also the possibility that too many programmes get started and compete for too few resources.

The lack of long-term industrial planning is also demonstrated by the large swings in naval shipbuilding workload. The MOD finds itself in a 'boom' and 'bust' cycle of shipbuilding that is difficult to break. The MOD is currently approaching the next 'boom' phase for naval shipbuilding (this last one occurred in the mid-1980s) with an industry that has diminished since the last 'boom'. After an anticipated peak around 2007–2010, the cycle will turn and the industry will be in a 'bust' phase. Only with a long-term view will the MOD be able to break this cycle. It will, however, require some difficult decisions with respect to timings of programmes.

In our view, there are a few key benefits to better long-term planning:

1. **Better financial planning.** Long-term planning will help the MOD (and Treasury) to have better forecasts of the financial implications of the DPA acquisition strategy. Levelling demand will help to level the spending on naval acquisition.
2. **Reduced cost and schedule risk.** Again, if the shipyards become overburdened or forced to expand too rapidly, the likely outcome is that ship programmes will slip or cost more than planned. It is not an issue of the programme failing outright, but rather one of efficiency.
3. **Help the DPA anticipate problems.** By looking at the demands on the yards, the DPA can independently anticipate problems and evaluate whether any remedy is needed. Right now, the DPA is

mostly in 'reactive' mode: The shipyards raise issues (possibly in a public forum), and the DPA must respond (or not). For better or worse, the DPA is the principal customer of the shipyards. Thus, it has a role in helping the shipyards be successful by moderating the demands on them.
4. **Aid in ending boom bust cycle.** By planning for the long term, the DPA can understand the implications of deferring programmes and can communicate these implications to other decisionmakers. Part of the reason for the near-term peak demand is that certain acquisitions (e.g., MARS) have been delayed so long that there is little schedule flexibility left if the Royal Navy needs to maintain this capability. By being able to show the effect of delay and deferral on the industrial base, the DPA can better advocate to decisionmakers potential industrial effects of their decisions. There seems to be little recognition of the consequences of these delays.

A strategic examination of the overall build programme with respect to the industrial impact should be done at least annually with an outlook of 10 to 15 years. The MOD needs to be disciplined in trying to balance its operational needs with industrial capacity and to be more proactive in balancing the workload it places on the shipyards. For better or worse, the MOD is the primary customer of the UK shipbuilding industry and will likely remain so indefinitely.

Define the Appropriate Role of the Offshore Industry
The UK offshore industry can contribute to the overall naval shipbuilding programme. The offshore industry brings several strengths to shipbuilding: experience working on collaborative programmes and alliance-type structures, a network of suppliers and subcontractors, experience with distributed design (engineering done in more than one location), broad programme management, and logistic skills. The offshore industry does have some weakness with respect to naval shipbuilding: lack of specific domain knowledge of military ships (systems, standards, etc.); limited experience with DPA proce-

dures, policies, and general programme; and limited steel plate and steel structure fabrication facilities.

The strengths of the offshore industry play well to the need for the naval shipbuilding industry to work in a more collaborative fashion. As we have stated earlier, better work-sharing between the shipyards will be necessary to meet the peak labour demand. The offshore industry might be the catalyst to make this work-sharing happen and function successfully. It is not likely, however, that the industry, in general, will feature strongly in direct fabrication; it might feature more prominently in assembly and integration.

Furthermore, the offshore industry's role, if any, will need to be matched to the product. The more commercial-like the product, the more significant role it can play. For the combatants, their role will be more limited.

Carefully Consider the Implications of Foreign Procurement of Complete Ships

Some observers have suggested that one possible (and potentially desirable) way to meet MOD demand would be to procure ships from foreign sources and thereby reduce the workload burden on the UK shipyards. Because our study focused on UK shipbuilding, we cannot comment on whether this approach is either feasible or desirable. However, one must be cognisant of several disadvantages to this approach:

- There are two economic implications to consider when exploring foreign procurement. First, there are domestic benefits to procuring ships in the United Kingdom. Money spent in the United Kingdom keeps workers employed and potentially off unemployment. The money that workers and companies spend will indirectly stimulate the economy further (e.g., service firms, housing providers, stores and shops). Second, there is an exchange-rate risk when contracting with a foreign firm. As has been seen with the JCA programme, these exchange rate swings

can lead to funding changes that are outside the control of the IPTs.[3]
- When considering procuring weapon systems from overseas, there is a concern that the United Kingdom will not have access to the latest or other advanced technologies. For example, countries generally do not export nuclear propulsion technology. Certain advanced sensors, missile systems, or stealth technologies may also be withheld from the export market. Another related concern is whether the United Kingdom could acquire solutions that are tailored to its specific requirements—as opposed to being limited to an off-the-shelf design.

Foreign procurement potentially leaves the United Kingdom susceptible to political interventions. For example, if a foreign government does not agree with a particular policy of the United Kingdom, it could delay or withhold the delivery of a vessel as a leverage point. Or such an inclined government could prevent the sale altogether. These types of problems are not unheard of; one needs only to look at the difficulty Taiwan has had in purchasing conventional submarines or the reversal of a frigate sale to Pakistan.

One might view this caution of foreign procurement as contradicting our earlier recommendation to relax the shipbuilding industrial policy. The difference between the two is the scale of items purchased. Our previous recommendation was to allow the yards to outsource work to foreign sources when (1) there was a need to reduce a labour peak, (2) the workforce was not available elsewhere in the United Kingdom, and (3) the workforce necessary would only be needed for a short period. We did not imply that entire vessels should be procured from foreign sources.

Labour Wage Pressures During Peak Demand

When demand exceeds supply in classic microeconomic theory, prices tend to rise. The MOD should be concerned that the labour demand

---

[3] See the explanation in National Audit Office (2002)

peak might cause upward pressure on shipbuilding wage rates. The link between wage escalation and skill shortages is disputed in the literature: Some argue that there is a strong correlation between the two[4], whereas others argue that the link is weak[5]. Nonetheless, the MOD should be aware that there might be higher-than-average wage-rate escalation in the shipbuilding industry during the labour peak and should develop its budgets accordingly.

Encourage Long-Term Investment Through Multi-Ship Contracts

As was noted in Chapter One, the naval shipyards have not modernised facilities during the past several years as the result of limited orders and a highly competitive market. One recent exception to this is the Type 45 programme. Because the initial contract was for six ships (a multi-ship buy), both BAE Systems and VT are incentivised to invest in facilities upgrades. Most notable is VT Shipbuilding, which has moved into a new facility in the Portsmouth Naval Shipyard. Only with longer-term contracts are the shipyards able to justify this type of major investment. If the Type 45 had been procured in annual lots of one to two ships, the investments made would be more difficult, if not impossible, to justify. The benefit to the MOD is that the shipbuilders will hopefully achieve greater efficiencies and pass reduced costs onto the defence ministry. It should be kept in mind, however, that such long-term contracts work better for mature designs and therefore may not always be appropriate for the first-of-class ship.

Consider the Feasibility of Competition in Light of the Industrial Base Constraints

The MOD prefers to compete programmes and production whenever possible and feels that competition achieves best value for money. However, the MOD should consider whether competition would yield better prices or result in a balanced allocation of work under an

---

[4] Frogner (2002, pp. 17–27).
[5] Robinson (1996).

environment in which there is high resource demands. One possibility in this environment is that there will be fewer potential bidders on subsequent programmes. This reduction would leave the later programmes with few bidders or no real competition. A second potential is that the bidders will actually take on more work than is feasible or optimal, thus creating potential later problems of cost growth or schedule slips. Finally, we have suggested that the shipyards will need to share workload to meet the labour peaks. Competition will make the shipyards less inclined to cooperate for fear of loosing some competitive advantage.

Explore the Advantages of Common Design Tools

As the MOD programme enfolds, the UK shipyards and firms will likely need to share both production and design resources to make the current plan feasible. One difficulty in sharing design resources is that each organisation has different three-dimensional CAD/CAM tools. Thus, interchanging data and working cooperatively on a common design will be difficult. The MOD might want to facilitate a discussion between the firms (and potentially include the CAD/CAM vendors) to explore whether the industry should adopt a common design tool, set of tools that are interoperable, or standards so that design work can be easily interchanged. Common design tools will also lead to common product models and databases, which will benefit the MOD in the long run with easier logistics support.

Conclusions

In this report, we have shown that the current shipbuilding plan of the MOD will be a challenge to the industry resources available in the United Kingdom. The overlap of several programmes in the next few years will result in a high demand for labour and facilities. Any potential shortfalls could result in cost increases and schedule delays. To mitigate these problems, we recommend that the MOD explore ways to level-load the demand through changes in programme timings, work with the DTI to encourage training in skills that are transferable to other industries, and encourage the use of more outsourcing. For the long term, the MOD needs to make industrial planning part of its acquisition process to better balance the demands it places on the industrial base and to break the boom-bust cycle.

APPENDIX A
Effects of Schedule Slippage on MOD Labour Demands

In Chapter Two, we examined the impact that the MOD's current and future shipbuilding programmes would have on the shipbuilding industrial base from the perspective of labour demand. Because there is uncertainty regarding the number of ships that the MOD will eventually order, we ran a number of scenarios to show the impact of differing ship order quantities on labour demand. At the end of the chapter, we then looked at a further scenario that attempted to level the MOD's labour demand over time to make it easier for the industrial base to meet any peak labour demands. The purpose of this appendix is to examine the effects of schedule slippage on MOD labour demands, since delayed programmes may cause increased and redistributed labour demands on the industrial base.

One of the implicit assumptions in all the scenarios we examined in Chapter Two was that all the shipbuilding programmes would run according to their planned schedules (i.e., there would be no delay in either the design and build schedules of any of the vessels). We made this assumption because it is reasonable for both the MOD and industry to use projected schedules to plan their labour demands, especially since it is difficult to anticipate in advance where unforeseen delays may occur.[1] However, we know from experience that delays do often occur and that they affect future labour projec-

---

[1] Additionally, we made this assumption since we were looking to model the MOD's *projected* demands—i.e., demands according to a set schedule.

tions. These delays often are associated with ships that are the first-of-class as both the government and industry modify requirements, reconcile design and production conflicts, and work out the problems associated with designing and producing a new type of ship. These delays often affect follow-on ships or other work in the shipyard, as industry readjusts its production schedules to take the changes into account. A careful read of recent NAO Major Projects Reports from the last few years shows the tendency for all defence procurement projects, and not only naval projects, to slip in schedule and grow in cost. Although the reasons for this schedule slippage and cost growth are outside of the scope of this report, we need to be aware of this trend and account for it in our analysis.

The modelling tools we developed for this study and used to forecast MOD demand in a variety of scenarios in Chapter Two are well suited to look at the impact of schedule slippage in the MOD's planned shipbuilding programme. The model can easily show the impact of delays in specific programmes, but it requires specific inputs defining precisely which programmes will be delayed, for how long, and how the total labour build hours will change because of the delays. Thus, in a complex shipbuilding programme such as the MOD is undertaking, one can imagine a large number of feasible schedule slippage scenarios, each with its own unique labour demand profiles. Because it is not feasible to examine each of these possibilities in detail, the aim of this appendix is to show, through a simple example, how the MOD could take schedule slippage into account when assessing its future labour demand.

It is also important to note that if specific ship schedules slip, there may be a number of knock-on effects within the industrial base. First, if a ship is delayed, the total labour hours required to build or design the ship will also likely increase. Schedule slippage will logically lead to cost (and labour hour) growth. There are many reasons for this. Project management must continue to staff the project for its entire duration; there may be rework required if designs change; production efficiencies may be lost should the schedule become less than optimal; or there may be additional facilities costs to account for the shipbuilding in a building hall or dry dock for longer than initially planned.

Second, if a ship is delayed, it may affect the design or production of other MOD ships being built during the same time frame. It is important to understand that, although ship classes are often managed by independent Integrated Project Teams within the Defence Procurement Agency, decisions or delays for one programme will almost certainly affect other programmes, should they be designed or produced by the same shipbuilder.

To illustrate potential effects on total labour demand of delays in design or production, we have devised a simple example, based on the baseline case presented in Chapter Two. This example looks at the impact of delays from a labour perspective only and does not consider facilities or labour supply limitations. For the purposes of this illustration, we took the MOD's prospective shipbuilding programme as we assumed in the Chapter Two baseline case and looked at the impacts on total labour demand if

- the production period for all MOD ships was extended by 24 months (from their original end date)
- total labour for each ship increased by 10 percent (from the base case).

We applied these assumptions only to the production of MOD ships (CVF, MARS, FSC, etc.); we excluded both the current refit programme, any export orders, and the nonrecurring design of any new vessels. Additionally, we did not apply these assumptions to the LSD(A) or LPD(R) programmes because we assumed that these ships were far enough into production as to not be delayed further.[2]

We realise that this specific scenario is unlikely—it is difficult to predict which, if any, programmes will slip as well as the duration and labour impact of any slip. Thus, for simplicity, we made the above assumptions as a straightforward example to examine the impact of

---

[2] We did, however, apply these assumptions to the Type 45 and *Astute* submarine programmes.

schedule slippage and labour growth.[3] This illustrates that it is possible to use the RAND labour projection model to do basic sensitivity analysis to examine a range of potential unforeseen shipbuilding schedules, and we encourage the MOD to specifically examine programme slippage scenarios that it feels warrant the most attention.

After changing the data inputs of the model to take into account our new assumptions, we calculated the MOD's future labour demand if all their programmes slipped by 24 months and the associated labour increased by 10 percent. We show the results in Figure A.1.

As one may expect with programme delays, the peak workload under this scenario has shifted later in time. It is now just over a year later than the peak labour demand expected under the current plan

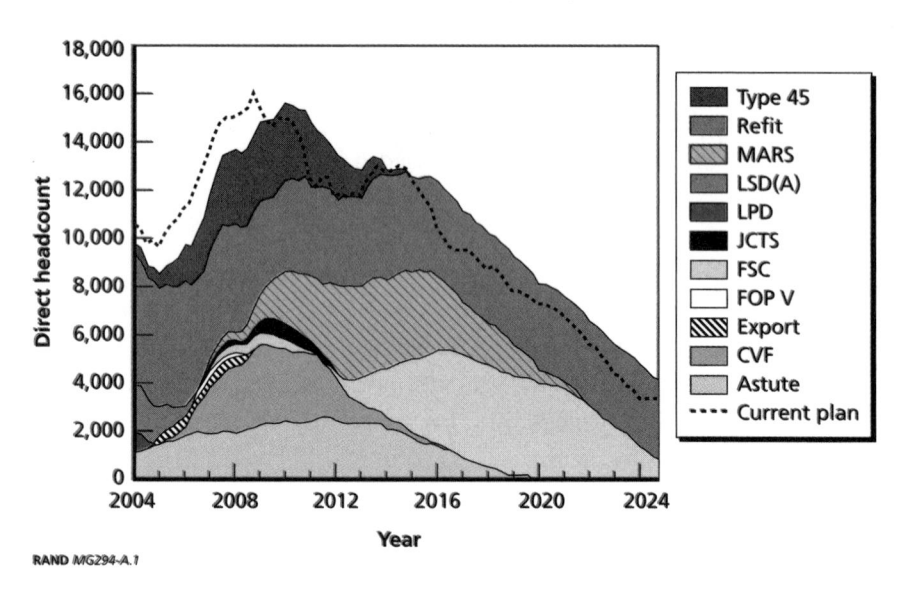

Figure A.1
Schedule-Slip Scenario Labour Projections by Programme

_____

[3] By making these assumptions, the RAND project team is not implying that the MOD shipbuilding programme schedule will slip. We are simply examining the impacts, *should* the programme slip to add further robustness to our results.

examined earlier and is slightly lower than the current plan's peak demand. However, it is also important to note that, because of schedule delays, there is significantly less labour demand for the next five years. This is compensated by an increased labour demand after the peak, again compared with the current plan of Chapter Two. Perhaps most substantially, because of an initial decrease in labour demand, the schedule-slip scenario actually requires the greatest absolute increase in workers from its low point in 2005 to its peak five years later. This absolute increase is greater than that needed should the programmes have encountered no schedule delays (i.e., the current plan in Chapter Two). Thus, although this scenario presents a later and slightly lower peak demand, the absolute increase in workers needed from the low to high point of demand is greater than if no delays existed.

In addition to looking at the overall MOD labour demand, we also were able to look at the impact of the programme delays on individual skill trades to see how these areas would be affected. We show the results in Figure A.2.

Looking at this figure, we see that there are no substantial differences from the comparable chart for the current plan examined in Chapter Two. There are, however, some minor differences. First, the labour demand curves are all shifted slightly later in time. Second, there is a greater initial decrease in outfitting skill level demands. Third, the schedule delays seem to have smoothed or removed the second demand peak that was originally seen in both the outfitting and structural trades in the current plan.

As we mentioned in Chapter Two, although this analysis is appropriate for the overall macro-level of labour demand, it is important to understand the micro-level demand impacts of schedule delays. Again, business sensitivities do not allow us to present the results at that level, but that does not diminish its importance to the MOD as we suspect that programme delays will have a much greater micro-level impact at the individual shipyard level, even if they do not greatly change the overall macro-demand.

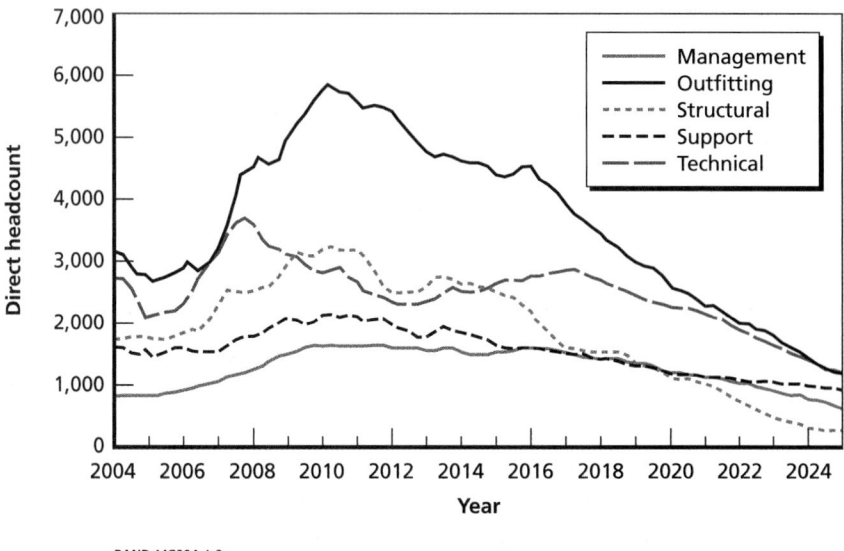

Figure A.2
Schedule-Slip Scenario Labour Projections by Skill Level

When one looks broadly at the impacts of schedule slipping on labour demand, it tends to move the peak demand into the future, and in the case of our scenario, slightly decrease it—similar to the overall effects of level-loading. This may lead some to question the need for level-loading, especially if excess demand or other circumstances will naturally level the demand without specific MOD planning. To further examine this issue, Figure A.3 shows the change in current MOD labour demand under the original current plan, our schedule slip scenario, and the level-loading plan presented at the end of Chapter Two.

The results of this comparison are extremely interesting. When comparing the level-loaded plan with the schedule slip scenario, it is clear that by selectively choosing which programmes to advance or delay, the MOD can decrease variation in labour demand much more

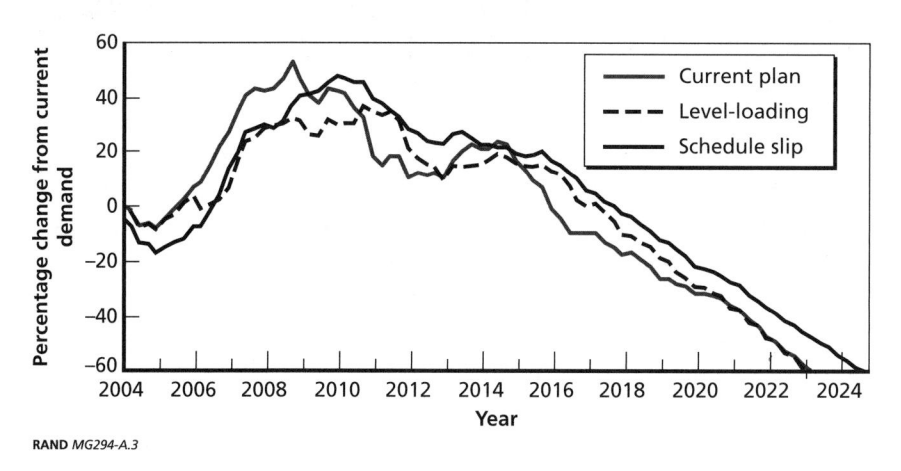

Figure A.3
Base Case, Level-Loaded, and Schedule Slip Demands Compared to Current MOD Demand

effectively than through slipping all of its programmes.[4] Although not conclusive, this should provide solid evidence of the benefits of level-loading, compared to either the current plan or allowing for a uniform schedule slip. By actively managing it's future shipbuilding programmes through level-loading and other mechanisms, the MOD will better be able to utilise the industrial base.

Thus, in this appendix we have looked briefly at the generic effects of schedule delay on overall MOD labour demand. As mentioned earlier, our aim was not to make any judgements of whether MOD shipbuilding programmes would be delayed and by how much, but rather to look more broadly at the impact of schedule delay on labour demand. It appears that schedule delays will push the labour demand peak farther into the future and may even slightly lower the peak demand. However, these delays may come at a price as

---

[4] The level-loaded percent line stays much closer to the zero percent change line in all early years compared with the schedule-slip percent line. This decreased variation should make it easier for the industrial base to better manage its workforce. The only time that the schedule slip scenario improves is after 2018, when the bulk of the current MOD shipbuilding programme is finished.

they may lead in the short-run to decreased labour demand followed by an intense growth of demand. Further, level-loading appears to be a better way to manage both peak labour demand and overall labour variation than hoping that schedule delays will smooth demand over time.

APPENDIX B
Ship Dimensions

Table B.1 lists the basic ship dimensions (rounded to the nearest metre) for those vessels described in Chapter Four. The data are taken from *The Royal Navy Handbook: The Definitive MoD Guide*.

Table B.1
Basic Ship Dimensions

	Length (metres)	Beam (metres)	Draught (metres)
AG[a] (*Argus*)	175	30	8
AO[b] (*Brambleleaf, Bayleaf,* and *Orangeleaf*)	171	26	11
Astute (SSN)	97	11	10
Bay class (LSD[A])	176	26	6
Fort class (AFS[c])	185	24	9
Invincible class (CVS)	209	34	7
LPD (*Albion* and *Bulwark*)	176	29	7
Rover class (AORL)	141	19	7
Type 23 (FFG)	133	16	5
Type 45 (DDG)	152	21	5

[a] Miscellaneous Auxiliary.
[b] Fleet Oiler.
[c] Combat Stores Ship.

APPENDIX C
Skill Breakout, by Management/Technical and Manufacturing Categories

The following page details job categories, broken out by specific skills (Table C.1).

Table C.1
Skill Breakout, by Management/Technical and Manufacturing Categories

Category/Subcategory	Specific Skill
Management/Technical	
General Management	Management
	Administration
	Marketing
	Purchasing
Technical	Design
	Drafting/CAD
	Engineering
	Estimating
	Planning
	Programme control/project management
Manufacturing	
Structure	Steelworker, plater, boilermaker
	Structure welder
	Shipwright/fitter
	Team leader, foreman, supervisor, progress control (fabrication)
Outfitting	Electrician, electrical technician, calibrator, instrument technician
	Hull insulator
	Joiner, carpenter
	Fibreglass reinforced pipe laminator
	Machinist, mechanical fitter/technician, fitter, turner
	Painter, caulker
	Pipe welder
	Piping/machinery insulator
	Sheet metal
	Team leader, foreman, supervisor, progress control (outfitting)
	Weapon systems
Direct Support	Rigger, stager, slingers, crane and lorry operators
	Service, support, cleaners, trade assistant, ancillary
	Stores, material control
	Quality assurance/control

References

Appledore International and University of Newcastle Upon Tyne, *Prospects for the UK Merchant Shipbuilding Industry*, prepared for the UK Department of Trade and Industry, June 2000.

Arena, Mark V., John F. Schank, and Megan Abbott, *The Shipbuilding and Force Structure Analysis Tool: A User's Guide*, Santa Monica, Calif., USA: RAND Corporation, MR-1743-NAVY, 2004.

Arena, Mark V., John Birkler, John F. Schank, Jessie L. Riposo, and Clifford Grammich, *Monitoring the Progress of Shipbuilding Programmes: How Can the DPA More Accurately Monitor Progress?* Santa Monica, Calif., USA: RAND Corporation, MG-235-MOD, forthcoming.

'Astute Sets Out on the Long Road to Recovery', *Jane's Navy International*, December 2003, pp. 28–30.

Bensaou, M., 'Portfolios of Buyer-Supplier Relationships', *Sloan Management Review*, Vol. 40, No. 4, Summer 1999, pp. 35–44.

Birkler, John, John F. Schank, Mark Arena, Giles K. Smith, and Gordon Lee, *The Royal Navy's New-Generation Type 45 Destroyer: Acquisition Options and Implications*, Santa Monica, Calif., USA: RAND Corporation, MR-1486-MOD, 2002.

Birkler, John, Denis Rushworth, James R. Chiesa, Hans Pung, Mark Arena, and John F. Schank, *Differences Between Military and Commercial Shipbuilding: Implications for the United Kingdom's Ministry of Defence*, Santa Monica, Calif., USA: RAND Corporation, MG-236-MOD, forthcoming.

Bradbury, John, 'North Sea Focus: Reinventing the North Sea', *Hart's E&P Net*, August 2000, www.eandpnet.com/ep/previous/0800/08northsea1.htm (last accessed November 2004).

Bruce, George, *Report on Skills Availability for the CVF Programme*, Thales Naval, 31 May 2002.

Burton, Anthony, *The Rise and Fall of British Shipbuilding*, London: Constable and Company, 1994.

Cavinato, Joseph L., and Ralph G. Kauffman, eds., *The Purchasing Handbook: A Guide for the Purchasing and Supply Professional*, 6th edition, McGraw-Hill, 1999.

Chaundy, David, *Skilled Trades and the Offshore Industry: A Skills Survey of Selected Construction Trades in Nova Scotia*, Atlantic Provinces Economic Council, October 2002.

Closhen, Cornelia, *Marine Engineering Labour Market Observatory*, Watford, Hertfordshire, UK: EMTA, 2002.

The Clyde Shipyards Task Force Report, Scottish Executive (Scotland Office), and UK Department of Trade and Industry, January 2002.

Department of Trade and Industry Engineering Industries Directorate, *Competitive Analysis of the UK Marine Equipment Sector*, March 2001.

Frogner, Mari Lind, *Labour Market Trends*, London: Office of National Statistics, January 2002.

Furness Enterprise Limited, *Reducing Strains on the Labour Resource Available for Warshipbuilding in the UK*, Burrow-in-Furness, Cumbria, UK, July 2003.

Gattorna, John, ed., *Strategic Supply Chain Alignment: Best Practices in Supply Chain Management*, Gower, UK: Andersen Consulting, 1998.

Hill, J. R., ed., *The Oxford Illustrated History of the Royal Navy*, Oxford, UK: Oxford University Press, 2002.

Jamieson, Alan G., *Ebb Tide in the British Maritime Industries: Change and Adaptation 1918–1990*, Exeter, Devon, UK: University of Exeter Press, 2003.

Johnman, Lewis, and Hugh Murphy, *British Shipbuilding and the State Since 1918: A Political Economy in Decline*, Exeter, Devon, UK: University of Exeter Press, 2002.

Laseter, Timothy M., *Balanced Sourcing: Cooperation and Competition in Supplier Relationships*, San Francisco, Calif., USA: Jossey-Bass, 1998.

Lloyd's Register, *Fairplay Ports and Terminals Guide*, 26 March 2004.

Merrow, Edward, 'Taking on the Cult of Mediocrity', *Upstream*, 23 May 2003.

Monczka, Robert, Robert Trent, and Robert Handfield, *Purchasing and Supply Chain Management*, 2nd edition, South-Western College Pubs, Thomson Learning, 2002.

Moore, Nancy Y., Laura H. Baldwin, Frank Camm, and Cynthia R. Cook, *Implementing Best Purchasing and Supply Management Practices: Lessons from Innovative Commercial Firms*, Santa Monica, Calif., USA: RAND Corporation, DB-334-AF, 2002.

'Mounts Bay Awaits Easter Launch Date', *Preview: The Journal of the Defence Procurement Agency*, March 2004.

National Audit Office, *Ministry of Defence Major Project Reports 2002: Report by the Comptroller and Auditor General*, HC 91 Session 2002-2003, December 2002.

Owen, Geoffrey, *From Empire to Europe: The Decline and Revival of British Industry Since the Second World War*, London: HarperCollins, 1999.

Robinson, Peter, in Jonathan Michie and John Grieve Smith, eds., *Creating Industrial Capacity: Towards Full Employment*, Oxford University Press, 1996, pp. 101–104.

Royal Navy, *The Royal Navy Handbook: The Definitive MoD Guide*, London: Conway Maritime Press, 2003.

Schank, John F., Hans Pung, Gordon Lee, Mark V. Arena, and John Birkler, *Outsourcing and Outfitting Practices: Implications for the Ministry of Defence Shipbuilding Programmes*, Santa Monica, Calif., USA: RAND Corporation, MG-198-MOD, forthcoming.

Shipbuilders & Shiprepairers Association, *Benchmarking UK Shipbuilding and Ship Repair*, Egham, Surrey, UK, January 2000.

Tang, Christopher S., 'Supplier Relationship Map', *International Journal of Logistics: Research and Applications*, Vol. 2, No. 1, 1999, pp. 39–56.

Todd, Daniel, and Michael Lindberg, *Navies and Shipbuilding Industries: The Strained Symbiosis*, Westport, Conn., USA: Praeger Publishers, 1996.

UK Offshore Operators Association, *Harnessing Talent and Technology, Economic Report*, London, 2000.

Walker, Fred M., *Song of the Clyde*, Edinburgh, UK: John Donald Publishers, 2001.

Winklareth, Robert J., *Naval Shipbuilders of the World: From the Age of Sail to Present Day*, London: Chatham Publishing, 2000.